ENVIRONMENTAL REMEDIATION TECHNOLOGIES, REGULATIONS AND SAFETY

GEOLOGIC CARBON DIOXIDE STORAGE ON FEDERAL LANDS

POTENTIAL AND ALLOCATIONS

ENVIRONMENTAL REMEDIATION TECHNOLOGIES, REGULATIONS AND SAFETY

Additional books in this series can be found on Nova's website under the Series tab.

Additional e-books in this series can be found on Nova's website under the e-book tab.

ENVIRONMENTAL REMEDIATION TECHNOLOGIES, REGULATIONS AND SAFETY

GEOLOGIC CARBON DIOXIDE STORAGE ON FEDERAL LANDS

POTENTIAL AND ALLOCATIONS

THERESA WISE
EDITOR

nova publishers
New York

Library of Congress Cataloging-in-Publication Data

ISBN: 978-1-63483-560-2

Published by Nova Science Publishers, Inc. † New York

Contents

PREFACE

As part of a global effort to assess the storage potential for captured carbon dioxide (CO_2), this book estimates and characterizes the storage potential that lies beneath some of the more than 400 million acres of Federal land available for lease. Estimated at between 126 to 375 billion metric tons of CO_2, the majority of this storage potential (about 85 percent) is located west of the Mississippi River, where most of the leasable Federal acreage (92 percent) is found. In assessing the potential for storage beneath Federal lands, the book addresses issues such as Federal control, location of emission sources, and pipeline rights-of-way, as well as laws and regulations relevant to Federal leasing.

In: Geologic Carbon Dioxide Storage … ISBN: 978-1-63483-560-2
Editor: Theresa Wise

Chapter 1

STORAGE OF CAPTURED CARBON DIOXIDE BENEATH FEDERAL LANDS*

Robert C. Dolence, Charles Kirschner, Christopher Babel, Patrick Findle and Steven Harouff

LIST OF ACRONYMS

BLM	Bureau of Land Management
BOR	Bureau of Reclamation
CCS	Carbon Capture and Sequestration (or Storage)
CERCLA	Comprehensive Environmental Response, Compensation, and Liability Act
COA	Conditions of Approval
DOD	Department of Defense
DOE	Department of Energy
DOI	Department of Interior
DOJ	Department of Justice
DOT	Department of Transportation
EA	Environmental Assessment
EIS	Environmental Impact Statement
EOR	Enhanced Oil Recovery

* This is an edited, reformatted and augmented version of a report, DOE/NETL-2009/1358, funded and issued by the National Energy Technology Laboratory, May 2009.

EPA	Environmental Protection Agency
EPACT05	Energy Policy Act of 2005
ESA	Endangered Species Act
FLPMA	Federal Land Policy and Management Act
FS	U.S. Forest Service
FWS	Fish and Wildlife Service
GIS	Geographic Information Systems
IOGCC	Interstate Oil and Gas Compact Commission
MGSC	Midwest Geological Sequestration Consortium
MRCSP	Midwest Regional Carbon Sequestration Partnership
MVA	Measure, Verification and Accounting
NATCARB	National Carbon Sequestration Database and Geographic Information System
NEPA	National Environmental Protection Act
NHPA	National Historic Preservation Act
NPS	National Park Service
PCOR	Plains CO_2 Reduction Partnership
PEIS	Provisional Environmental Impact Statement
RCRA	Resource Conservation and Recovery Act
RCSP	Regional Carbon Sequestration Partnership
ROW	Right of Way
SDWA	Safe Drinking Water Act
SECARB	Southeast Regional Carbon Sequestration Partnership
SWP	Southwest Partnership CO_2 Sequestration
TUP	Temporary Use Permits
USDA	U.S. Department of Agriculture
USDW	U.S. Drinking Water
USGS	U.S. Geological Survey
UIC	Underground Injection Control
WESTCARB	West Coast Regional Carbon Sequestration Partnership

EXECUTIVE SUMMARY

A global effort is underway to assess storage potential for captured carbon dioxide (CO_2). In North America, the *Carbon Sequestration Atlas of the United States and Canada* provided an initial assessment. The analysis presented in this report was done to assess the storage potential beneath Federal lands and further the effort undertaken in the atlas by defining a resource potential beneath a specific category of land. Also considered in this analysis was the location of potential CO_2 point sources that might utilize Federal lands for storage, pipeline Right-Of-Way (ROW), and wells located on or near Federal land. Relevant laws, regulations, and legislation at the Federal and State level are also summarized.

A significant portion of Federal land is unavailable for leasing due to administrative, statutory, and executive orders. Examples include national parks and lands owned by the Department of Defense (DOD), the Department of Justice (DOJ), and other agencies. These limitations render 44 percent of all Federal acreage unavailable for lease. Remaining Federal lands, totaling 400,730,534 acres, are available for lease (Figure ES-1). The Bureau of Land Management (BLM) controls 59 percent of this acreage and the U.S. Forest Service (FS) controls another 40 percent with the balance managed by the Department of Energy (DOE) and the Bureau of Reclamation (BOR). All of the BLM acreage and 80 percent of FS acreage is west of the Mississippi.

The storage resource beneath Federal lands ranges between 126 and 375 billion metric tons (Table ES-1). Since the vast majority of Federal lands are west of the Mississippi, it follows that the majority of storage potential beneath these lands are also located in the western half of the Nation. Of the estimated storage potential beneath Federal land in the United States, 68 percent can be found in the stratigraphy of Montana, Wyoming, and the Dakotas.

Conversely, the majority of CO_2 point sources in the United States are found east of the Mississippi. Federal lands are not as contiguous in the east as they are in the west; however, there is some storage potential available for consideration, the majority of which is found in the stratigraphy of the Gulf Coast states and Arkansas.

Saline formations account for between 71 and 90 percent of the total carbon storage potential beneath Federal lands. Oil and gas reservoirs provide between 9 and 25 percent of Federal land storage potential. Unmineable coal seams provide a further 1 to 3 percent. Whereas saline formations and unmineable coal seam resource estimates present a low and high range of potential, the storage potential estimate for oil and gas reservoirs is a single quantity: 32 billion metric tons. This reflects the higher level of knowledge operators have about these reservoirs due to oil and gas activity. It also represents a good opportunity for additional recovery of an important energy resource through Enhanced Oil Recovery (EOR) operations.

In the interest of furthering Carbon Capture and Sequestration (CCS) efforts, Federal lands present a unique advantage over those that are privately owned: single ownership of large, continuous acreage tracts. Negotiating with a single landowner to secure the rights to extensive continuous parcels of land can provide cost and project timeline advantages, not only for potential future operations but also for early large-scale demonstration projects that will help accelerate commercial deployment of CCS technology.

Federal lands, however, come with conditions, stipulations, covenants, restrictions, and other obligations associated with the protection of our Nation's environmental, historic, and cultural treasures. Compliance with National Environmental Protection Act (NEPA), Endangered Species Act (ESA), and National Historic Preservation Act (NHPA) can be daunting, but it is necessary to capitalize on the extensive, publicly held acreage available for development. The initial impact of such conditions will be in lease accessibility. Many leases carry one or more restrictions, for example, seasonal limitations on activity for the protection of certain species and ecosystems. These regulations will impact pipeline ROWs and most likely CO_2 storage operations, as well.

The single most important challenge to utilizing Federal lands for storage of captured CO_2 is regulatory. Experience from oil and gas operations provides an important guide and basis for development of regulations and procedures for CO_2 storage operations. A regulatory framework will help to define expectations for the operator, regulator, and insurer. The EPA and the states of Washington and Wyoming have provided an important yet only initial effort in the regulatory arena.

Storage of captured CO_2 is expected to be permanent. Which individual(s), organization(s), or other entities will take responsibility for long-term liability once injection operations have been completed has yet to be determined. Prospective storage field operators will want to know that their obligations can come to an end upon successful completion of their work. Resolution of this situation regarding long-term liabilities is critical to successful deployment of CCS technology. It is imperative that this work, along with suitable regulations, be continued and resolved so that CCS operations can begin when supporting technologies are ready for large-scale demonstration and commercial deployment.

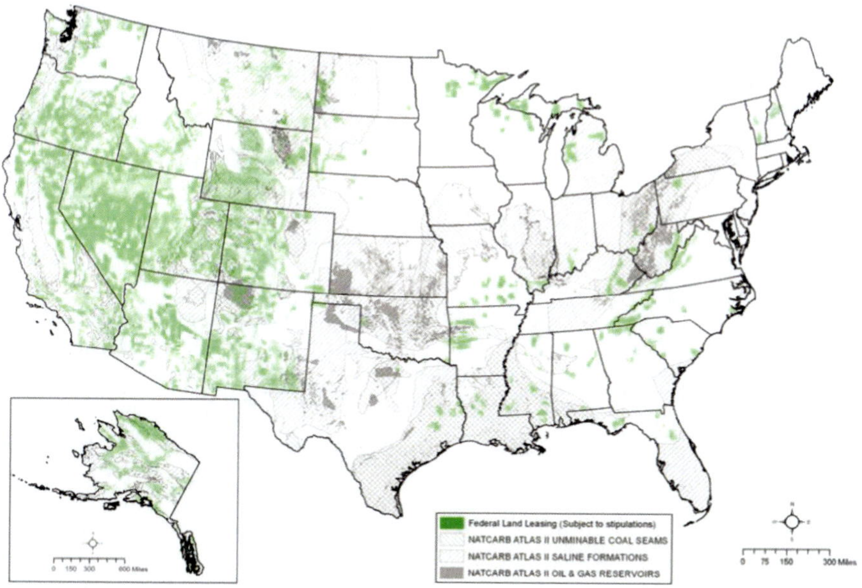

Figure ES-1. Distribution of Leasable Federal Lands and Potential Storage Capacity.

Table ES-1. Summary Points for CO_2 Storage Beneath Federal Lands

CO_2 Storage Potential by Type of Reservoir	**Storage Potential – Billion Metric Tons**		**Comments**
	Low Estimate	**High Estimate**	
Federal Land	126	375	About 5% of national storage capacity[1]
Saline Reservoir	90	337	Half of storage potential in Wyoming of Big Sky region
Oil & Gas Reservoir	32	32	80% of storage potential in the Dakotas, Montana and Wyoming of PCOR region
Unmineable Coal Seams	4	6	3/4 of storage potential in Gulf Coast of SECARB region
Advantages	Single Ownership, good potential to convey surface and subsurface rights.		
	Potential to lease large tracks of land.		
	Establishment of ROW corridors.		
Challenges	65% of emissions are east of the Mississippi River.		
	83% to 86% of storage capacity west of Mississippi River.		

Table ES-1. (Continued)

CO_2 Storage Potential by Type of Reservoir	Storage Potential – Billion Metric Tons		Comments
	Low Estimate	High Estimate	
	Environmental qualifications: NEPA, ESA, NHPA. Present conditions of approval (COA) in gaining a lease.		
	No current lease terms and conditions address the ownership of the pore space and storage of CO_2 on the Federal lands.		
	Existing laws and regulations have analogous requirements for projects, such as those for natural gas, but they require modification and/or additions to accommodate CO_2 projects.		
	Short-term and long-term liabilities must be clearly defined before permanent CO_2 storage beneath Federal lands can be broadly deployed.		
	Without a transfer of liability upon conclusion of active injection operations, prior to long-term storage, many believe that potential operators will not be willing to enter into a situation from which there is no release.		
	Long-term storage of CO_2 beneath Federal lands will require long-term monitoring for potential subsurface and surface leakage.		
	Although CO_2 is listed as a Class 2.2 (non-flammable gas) hazardous material under DOT regulations (49 C.F.R. §172.101), the agency applies nearly the same safety requirements to CO_2 pipelines as it does to pipelines carrying such hazardous liquids as crude oil, gasoline, and anhydrous ammonia (49 C.F.R. § 195).		

Federal lands present an excellent resource for the storage of captured CO2. Once the technologies are in place and regulatory details are sorted out, this resource can be developed.

1.0. Carbon Dioxide Storage Potential Beneath Federal Lands

The National Energy Technology Laboratory (NETL) has lead the effort in evaluating the geologic storage potential for captured carbon dioxide in the United States, the downstream end of the carbon capture and storage chain. Through the efforts of NETL's Regional Carbon Sequestration Partnership program, a carbon sequestration atlas was published in 2007. This atlas provided estimated storage potential by geologic reservoir, saline, oil and gas or unmineable coal seam, and illustrated its distribution. A second edition of the atlas was published in November 2008.

Access to a carbon dioxide storage reservoir requires control of surface and subsurface rights; it requires the assembly of a block of acreage that will assure surface access for all operations and ownership of the subsurface pore space. The Federal Government is the largest land owner and some of this acreage is available for commercial development. The work presented in this report is an initial effort to estimate the storage potential beneath Federal land. Which Federal agency or department controls this acreage, emissions sources that may utilize this land and pipeline right-of-ways are describe in the first section. A brief discussion of relevant laws and regulations that impact leasing Federal acreage is second followed by conclusions that tie everything together.

1.1. Introduction

Carbon capture and storage (CCS) is considered an essential technology for controlling the emission of CO_2 into the atmosphere. Conservation, renewable energy, and improvements in the efficiency of power plants, automobiles, and other energy consumption devices are important first steps in any greenhouse gas emissions mitigation effort. But those approaches cannot deliver the level of emissions reduction needed to stabilize the concentrations of greenhouse gases in the atmosphere—especially against a growing global demand for energy. Technological approaches are needed that are effective in reducing atmospheric greenhouse gas concentrations yet, at the same time, have little or no negative impacts on energy use and economic growth and prosperity. CCS efforts hold great promise as a greenhouse gas reduction strategy.

Storage is a critical component in CCS technology. Subsurface or geologic storage presents the greatest potential for sequestration of captured CO_2. The work presented in this report is an initial assessment of geologic storage potential beneath a specific category of land, Federal land. Commercial use of Federal lands is a common activity for, among other activities, oil and gas production, mining, logging, recreation and grazing for livestock. Not only is the Federal Government the largest single landowner in the United States, it maintains extensive continuous land holdings. The volume of CO_2 expected to be captured and sequestered will require extensive subsurface storage space. Under these circumstances, the question arises whether sufficient Federal acreage could be leased for a large CO_2 storage field operation without the complication of interfacing with multiple landowners. Federal lands are worth considering as an option for demonstration and/or early deployment sequestration projects.

Acquiring the lease is only part of the challenge, however. Many questions must be answered regarding the volume of CO_2 captured by the source, the distance of the source to the proposed storage field (sink), right-of-way (ROW) for the pipeline between source and sink, and regulations needed at the State and Federal level to govern this type of operation.

Transportation and injection of CO_2 into underground strata is familiar technology and has long been a standard part of enhanced oil recovery (EOR). This activity provides a nucleus of technical data, information and associated regulations for further development of CO_2 storage fields and the necessary regulatory framework. The technologies necessary for capturing CO_2 from stationary sources in combination with storing it permanently underground are approaching commercial viability. Now, a variety of non-technical issues associated with application of this technology needs to be addressed in preparation for demonstration and early deployment sequestration projects. Among these issues are site selection, gaining rights to use private and/or public land for CO_2 transportation and storage, and the establishment of physical and regulatory infrastructures that can support carbon management on a large scale.

This report provides an initial estimate of the storage potential that lies beneath leasable Federal land for sequestering captured carbon dioxide. It also reviews existing and proposed laws and regulations for their potential impact on transportation, injection, and storage of CO_2 beneath Federal lands. Securing a lease for oil and gas exploration and production is an established practice. Doing so for injection of CO_2 for storage beneath Federal lands is not. Any additional requirements or considerations regarding CO_2 injection operations and storage needs to be sorted out before a lease is issued.

Data and information required to map the various elements needed for this study was gathered from various sources. The National Carbon Sequestration Database and Geographic Information System (NATCARB)[2] is the primary source for this study. NATCARB gathers the data submitted by the U.S. Department of Energy's Regional Carbon Sequestration Partnerships (RCSP) regarding sources and geologic sinks and merges the data into a national database. Federal land records were acquired from the U.S. Geological Survey (USGS). This Geographic Information System (GIS) database provided information on land boundaries and the agency that manages each land tract. Pipeline ROW data for CO_2 pipelines were provided by NATCARB. GIS data for natural gas and petroleum pipelines were obtained through the Ventyx Energy Velocity database (February 2008)[3] and the National Pipeline Mapping System database (July 2005).[4] Well data was acquired from various State agencies. The software platform used in this study is ESRI ArcViewGIS 9.0 (see Appendix A for more details on management and modeling of GIS data).

This study was conducted in conjunction with the *Carbon Sequestration Atlas of the United States and Canada*, which was prepared with technical contributions from each member of the RCSP Initiative. The sources, infrastructure, and geologic sinks corresponding to Federal land tracts were evaluated for each RCSP. The associated boundaries for each of the seven RCSPs (Figure 1) and the Northeast (not an RCSP region) were collected and, as a basic modeling domain, used as initial analysis areas. The RSCP regions are (Figure 1):

Figure 1. Regional Carbon Sequestration Partnerships.[5]

- Big Sky Carbon Sequestration Partnership (Big Sky)[6]
- Midwest Geological Sequestration Consortium (MGSC)[7]
- Midwest Regional Carbon Sequestration Partnership (MRCSP)[8]
- Plains CO_2 Reduction Partnership (PCOR)[9]
- Southeast Regional Carbon Sequestration Partnership (SECARB)[10]
- Southwest Partnership CO_2 Sequestration (SWP)[11]
- West Coast Regional Carbon Sequestration Partnership (WESTCARB)[12]

The boundaries of two partnerships, PCOR and WESTCARB, include land areas in Canada; however, only the lands within the United States were used for this study. The Northeast region of the United States does not have any assessed storage potential and these states did not join any of the NATCARB[13] partnerships. The Northeast states—Connecticut, Delaware, Maine, Massachusetts, New Hampshire, New Jersey, Rhode Island, and Vermont—were grouped to form a Northeast region for the study. Hawaii recently joined WESTCARB but was not included in this study because CO_2 storage in basalt is still in the experimental stage.

Storage potential for each of the prospective reservoir types (saline formations, oil and natural gas reservoirs, and unmineable coal seams) is based on the data prepared for the second edition of the atlas (2008) and represents the updated estimates calculated by each of the partnerships. With this basic information, this report provides an initial estimate of storage potential beneath Federal land; this information has been incorporated into the second edition of the atlas.

The evaluation of CO_2 emission sources was limited within this study to those sited within 100 miles (161 km) of Federal land. ROW and well data presented here are also subject to the 100- mile (161 km) limit. The distance over which to transport captured CO_2 for sequestration is an economic decision based on many variables outside the scope of this report. The purpose of this report is to provide an assessment of potential storage capacity of CO_2 beneath Federal lands as well as describe the relevant regulations, or regulatory gaps that will affect utilization of Federal lands for storage of captured carbon dioxide.

1.2. Federal Land Groups

This study used an existing USGS spatial shapefile[14] to identify lands owned and/or administered by the Federal Government. The USGS dataset categorizes Federal landholdings under 65 separate government bodies. To obtain a manageable description of Federal holdings, the 65 separate entities were organized into nine Federal "land groups" according to common Department or Agency (Table 1):

1) Department of Defense (DOD)
2) Department of Energy (DOE)
3) Bureau of Land Management (BLM)

4) Bureau of Reclamation (BOR)
5) Fish and Wildlife Service (FWS)
6) National Park Service (NPS)
7) Department of Justice (DOJ)
8) Other Federal agencies
9) U.S. Forest Service (FS) and other U.S. Department of Agriculture (USDA)

Distribution of Federal acreage by agency or bureau is illustrated in Figure 2. A list of agencies and bureaus identified by USGS as they were organized for this study is provided in Appendix B. (Only surface ownership is identified in the USGS National Atlas[15]; therefore split estates—in which the Federal surface owner does nor hold the mineral and/or pore space rights were excluded from the study.)

Not all Federal land is available for leasing. Administrative, statutory, and executive orders prohibit the lease of an estimated 44 percent of Federal land holdings in the United States. By comparing the 65 individual USGS entities with the nine Federal land categorizations found in *Inventory of Onshore Federal Oil and Natural Gas Resources and Restrictions to Their Development: Phase III Inventory*[16], leasable and non-leasable portions of Federal land were organized into the following three general access categories:

1) No leasing (some exceptions may exist)
2) Leasing (subject to stipulations)
3) Leasing status unknown

Note that the assignment of a category to a specific land tract does not equate to a definitive lease decision for that tract. Each parcel is protected by site-specific provisions in its enacted legislation that identify the specific land-use policies of that tract.[17] Thus, some tracts generally identified as unleasable may contain areas that could be leased for limited use, and some tracts generally identified as leasable may contain areas that are off limits to public use. Distribution of Federal acreage by one of three general access categories is illustrated in Figure 3.

Table 1. Federal Land Group Acreage Within U.S. Region (Acres x 1,000)

		Group 1	Group 2	Group 3	Group 4	Group 5	Group 6	Group 7	Group 8	Group 9
RCSP REGION	REGION / GROUP TOTAL	DOD	DOE	BLM	BOR	FWS	NPS	DOJ	Other Federal Agencies	FS and other USDA
PCOR*	34,587	2,046	0	7,430	158	2,103	1,236	0.5	0	21,614
MGSC*	2,364	481	14	0	0	151	51	1	152	1,512
MRCSP*	11,733	718	12	0	0	234	413	0	5	10,351
WESTCARB*	448,738	12,136	1,205	184,628	892	95,713	67,636	0	0.5	86,530
SECARB*	33,182	4,329	270	0	54	3,300	3,115	3	687	21,424
SWP*	106,350	7,128	23	47,981	469	913	4,406	1	33	45,397
Big Sky*	81,330	291	574	29,494	1,167	319	3,917	0	0	45,567
Northeast*	1,957	146	0	0	0	201	88	0.5	0	1,522
Total	720,240	27,275	2,098	269,533	2,738	102,934	80,862	6	878	233,916

* PCOR (Plains CO_2 Reduction Partnership), MGSC (Midwest Geological Sequestration Consortium), MRCSP (Midwest Regional Carbon Sequestration Partnership) WESTCARB (West Coast Regional Carbon Sequestration Partnership), SECARB (Southeast Regional Carbon Sequestration Partnership), SWP (Southwest Regional Carbon Sequestration Partnership), Big Sky (Big Sky Carbon Sequestration Partnership), and Northeast which did not join any of the NATCARB partnerships.

(Appendix C contains a list of agencies and bureaus by their basic lease access status). It can be seen in Table 2 and confirmed by the map in Figure 3 about 85 percent of leasable Federal land is essentially west of a line running from eastern New Mexico up through the western Dakotas. This concentration of acreage lies in three partnerships (defined in the next section), WESTCARB, SWP and Big Sky. East of the Mississippi, most of the Federal acreage, U.S. Forest Service land, is present in the SECARB partnership.

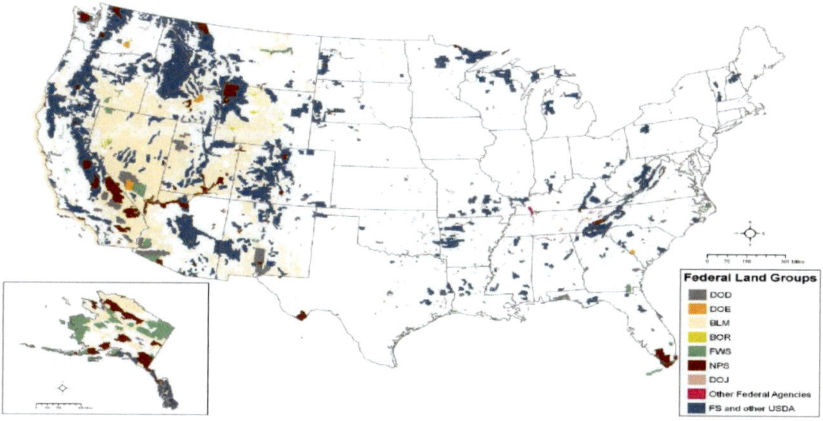

Figure 2. Federal Land Groups.

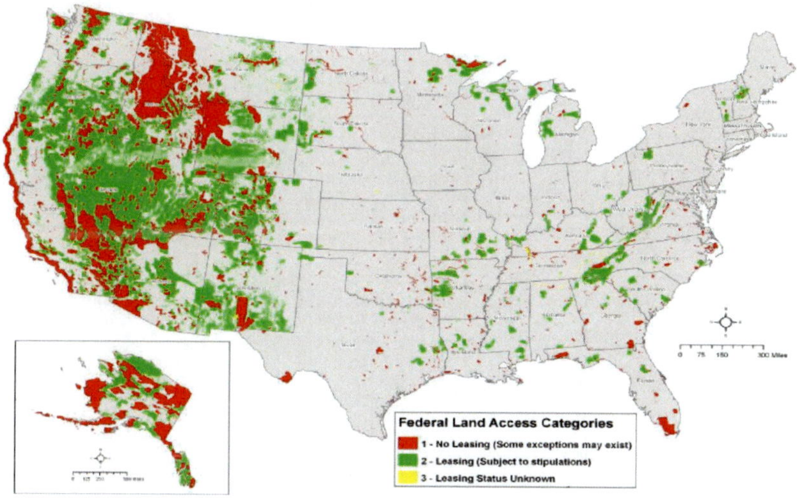

Figure 3. Federal land Access Categories.

Table 2. Leasable Federal Land Group Acreage Within U.S. Region (Acres x 1,000)

		Group 1	Group 2	Group 3	Group 4	Group 5	Group 6	Group 7	Group 8	Group 9
RCSP REGION	REGION / GROUP TOTAL	DOD	DOE	BLM	BOR	FWS	NPS	DOJ	Other Federal Agencies	FS and other USDA
PCOR	26,444	0	0	7,155	158	0	0	0.5	0	19,130
MGSC	1,484	0	14	0	0	0	0	1	0	1,468
MRCSP	10,033	0	12	0	0	0	0	0	0	10,021
WESTCARB	229,607	0	1,205	158,883	890	0	0	0	0.5	68,629
SECARB	20,915	0	270	0	54	0	0	3	0	20,588
SWP	81,082	0	23	41,870	469	0	0	1	32	38,687
Big Sky	29,835	0	574	27,988	1,167	0	0	0	0	106
Northeast	1,331	0	0	0	0	0	0	0.5	0	1,331
Total	400,731	0	2,098	235,896	2,738	0	0	6	33	159,959

1.3. CO_2 Point Sources

According to the *Carbon Sequestration Atlas*, “electricity generation” and “industrial” sectors account for the majority of CO_2 emissions from stationary sources—86 percent and 4.6 percent, respectively. Remaining emissions from stationary sources are split among agriculture processing, cement manufacturing, ethanol production, fertilizer production, petroleum and natural gas processing, petroleum refineries, and the chemical industry (see Table 3).

Nationally, the locations of 4,304 stationary sources were identified on and within 100 miles (161 km) of leasable Federal land in the United States. Of these, approximately 2,384 emit greater than 10,000 metric tons of CO_2 per year (see Figure 4). A single source may be within 100 miles (161 km) of several Federal land parcels, each controlled by different agencies. CO_2 source data was provided by NATCARB (see Appendix A).

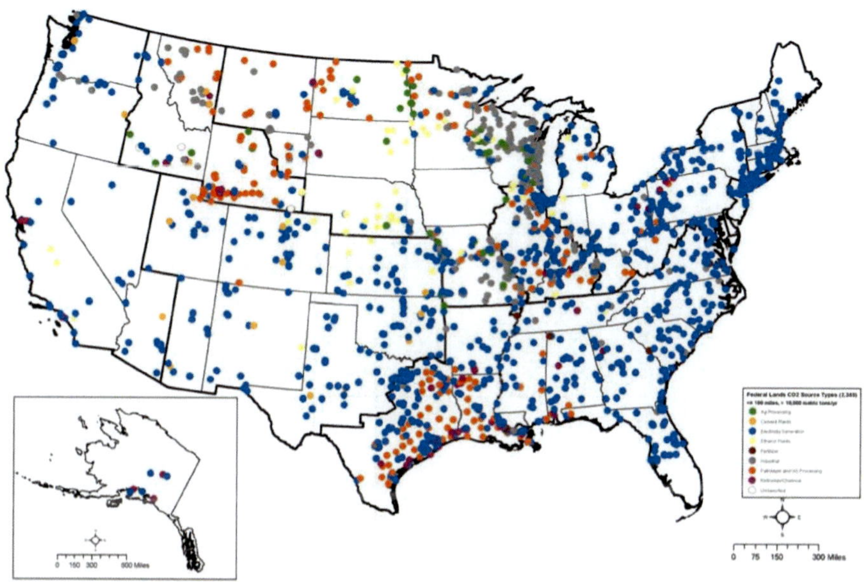

Figure 4. CO_2 Emitting Sources Producing > 10,000 Metric Tons of CO_2 Per Year and Located on or Within 100 Miles of Leasable Federal Land.

Table 3. Percent of CO_2 Emissions of Each Source Type in Each Region

Source	WESTCARB	SWP	BIG SKY	PCOR	MGSC	MRCSP	SECARB	NE
Ethanol Plants	0%	1%	0%	4%	1%	0%	0%	0%
Cement Plants	3%	4%	3%	1%	0%	0%	0%	0%
Ag Processing	0%	0%	2%	2%	0%	0%	0%	0%
Electricity Generation	87%	91%	77%	79%	90%	90%	89%	98%
Fertilizer	0%	0%	0%	0%	0%	0%	1%	0%
Industrial	1%	2%	8%	11%	6%	8%	1%	0%
Petroleum & NG	0%	1%	7%	2%	0%	0%	1%	0%
Refineries / Chemical	8%	1%	3%	1%	2%	2%	8%	2%
Unclassified	0%	0%	1%	0%	0%	0%	0%	0%

Table 4. Facilities >10,000 Mt/Yr CO_2 On and Within 100 Miles of Leasable Federal Land

		Group 1 Total	Group 2 Total	Group 3 Total	Group 4 Total	Group 5 Total	Group 6 Total	Group 7 Total	Group 8 Total	Group 9 Total
RCSP REGION	REGION TOTAL	DOD	DOE	BLM	BOR	FWS	NPS	DOJ	Other Federal Agencies	FS and other USDA
PCOR	667	0	0	143	157	0	0	55	0	583
MGSC	399	0	222	0	0	0	0	90	0	348
MRCSP	580	0	128	0	0	0	0	0	0	454
WESTCARB	135	0	13	113	81	0	0	0	16	126
SECARB	785	0	52	0	140	0	0	85	23	707
SWP	362	0	3	202	245	0	0	29	14	290
Big Sky	231	0	13	230	197	0	0	0	0	21
Northeast	169	0	0	0	0	0	0	117	0	86

Table 5. CO_2 Emissions (million metric tons (MMt/Yr))
Produced On and Within 100 Miles of Leasable Federal Land

		Group 1 Total	Group 2 Total	Group 3 Total	Group 4 Total	Group 5 Total	Group 6 Total	Group 7 Total	Group 8 Total	Group 9 Total
RCSP REGION	REGION TOTAL	DOD	DOE	BLM	BOR	FWS	NPS	DOJ	Other Federal Agencies	FS and other USDA
PCOR	500	0	0	131	126	0	0	64	0	427
MGSC	660	0	285	0	0	0	0	165	0	677
MRCSP	800	0	61	0	0	0	0	0	0	741
WESTCARB	200	0	17	170	101	0	0	0	22	195
SECARB	1,070	0	91	0	150	0	0	144	13	1,020
SWP	440	0	1	315	309	0	0	30	19	406
Big Sky	160	0	3	156	106	0	0	0	0	5
Northeast	82	0	0	0	0	0	0	58	0	40

Some or none of the Federal land parcels may be suitable for CO_2 storage. As discussed later, suitability of Federal land for CO_2 storage will be determined by a combination of geologic and geographic factors and the ability to obtain a lease permitting access to the prospective acreage.

The number of facilities emitting more than 10,000 metric tons per year and the volume of emissions from these facilities are posted in Table 4 and Table 5 respectively. Since any particular facility and its emissions can be within 100 miles of more than one Federal land group within or across RCSP boundaries, Land Group totals do not equal Region Totals and Region Totals do not equal national totals. In spite of this modeling dilemma, some general trends can be seen in these two tables. While just over half of the facilities are east of the Mississippi River (MGCS, MRCSP and SECARB), these facilities account for almost two-thirds of the emissions. The contradiction here is that less than 10 percent of Federal acreage is found east of the Mississippi River (Table 2 and Figure 3) while the WESTCARB partnership contains just over half of the Federal acreage yet contributes next to the lowest volume of emissions.

1.4. Existing CO_2 Infrastructure

Within the contiguous states, an extensive infrastructure exists that might be used for the sequestration of CO_2 beneath Federal land. This includes oil and gas wells, currently operating CO_2 pipelines, and existing ROWs.

There are millions of active and inactive oil and gas wells in place throughout the United States. Approximately 152,000 wells are located directly on Federal leasable acreage; approximately 2.58 million wells lie within 100 miles (161 km) of leasable Federal land (see Tables D-7 and D8 in Appendix D). Wells on or immediately offsetting Federal acreage could be used for the injection of CO_2. However, as with well storage anywhere, caution must be exercised. These wells, depending on their depth, may present a breach in cap rock, providing a possible path of CO_2 migration to overlying horizons or to the surface. The depth, age, and mechanical integrity of these wells must be carefully evaluated by any potential storage field project developer. Wells that cannot be used for injection purposes, either because of characterization or distance from the Federal land under consideration, still represent data points to interpret and map the subsurface beneath and around Federal acreage.

The majority of the CO_2 pipeline network now in place is located in Texas, New Mexico, and Wyoming. This discontinuous network of pipelines stretches over approximately 4,000 miles through nine states (Figure 5).[18] These pipelines can provide a nucleus for development of a larger network. This work is currently underway in Mississippi, Louisiana, and the Texas Gulf Coast.

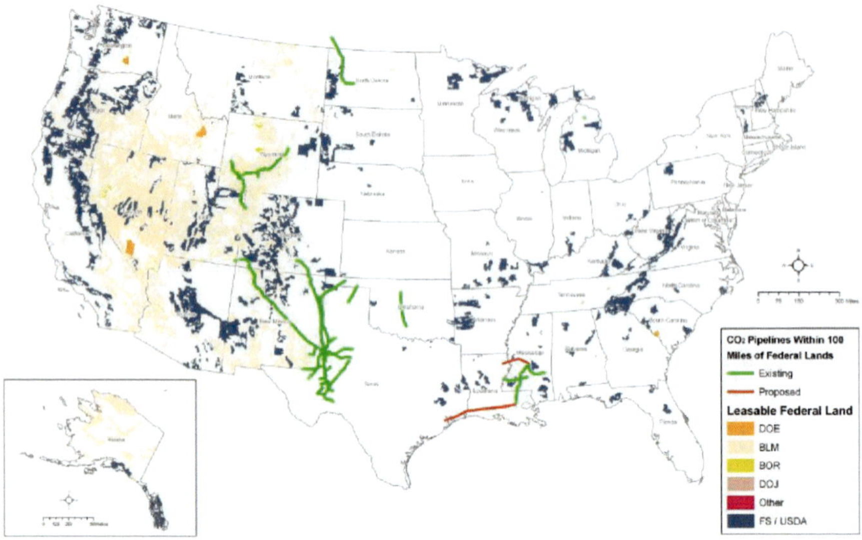

Figure 5. Existing and Proposed CO_2 Pipelines and Leasable Federal Land.

A vast and intricate network of natural gas and petroleum pipelines currently crisscross the Nation (Figure 6). These pipelines cannot transport CO_2 (unless retrofitted); however, the ROWs established for them may provide sufficient space for the construction of CO_2 pipelines.

The process of establishing a ROW can be cumbersome. Almost any route will cross many different tracts of property between the Federal land boundary and the CO_2 source point. Each property owner must be contacted and negotiations conducted to obtain the ROW. During this process, some or all property owners may be demanding regarding compensation and/or information on the project. Access to established ROWs would allow CO_2 pipelines to parallel the current natural gas and petroleum network. Utilizing existing pipeline ROWs for the construction of CO_2 pipelines will help minimize the time and cost of developing a CO_2 pipeline infrastructure.

The pipeline network illustrated in Figure 6 is thin in the western states where the majority of Federal land is present. Utilizing established ROWs may facilitate transportation of captured CO_2 across regional distances for storage or EOR opportunities. The existing CO_2 pipeline network will help facilitate use of Federal lands in this area.

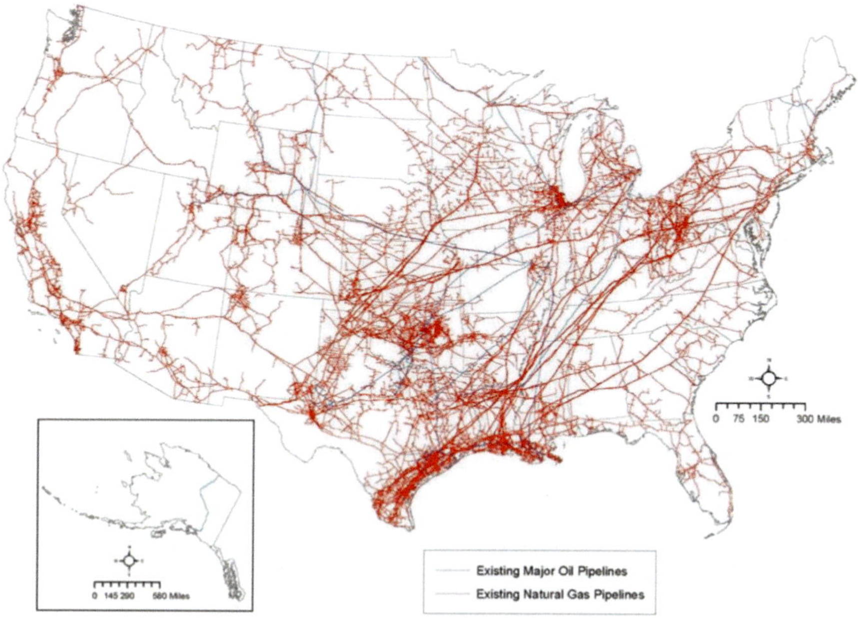

Figure 6. Existing Natural Gas and Petroleum Pipelines.

1.5. Geologic Storage Formations and CO_2 Capacity

Multiple geologic horizons in the subsurface have the potential for sequestration of CO_2 beneath Federal land. Within the geologic column, three types of storage space are considered suitable: saline formations, oil and gas reservoirs, and unmineable coal seams. Distribution of these resources is illustrated in Figure 7.

Deep saline formations present an immense storage potential for CO_2, considerably more than the potential provided by oil and gas reservoirs and unmineable coal seams combined. Saline formations are defined as permeable and porous rock containing water with total dissolved solids greater than

10,000 mg/L. Under standard gradients, a depth of approximately 2,800 feet will provide appropriate temperature and pressure conditions to keep the CO_2 in a supercritical phase.[19]

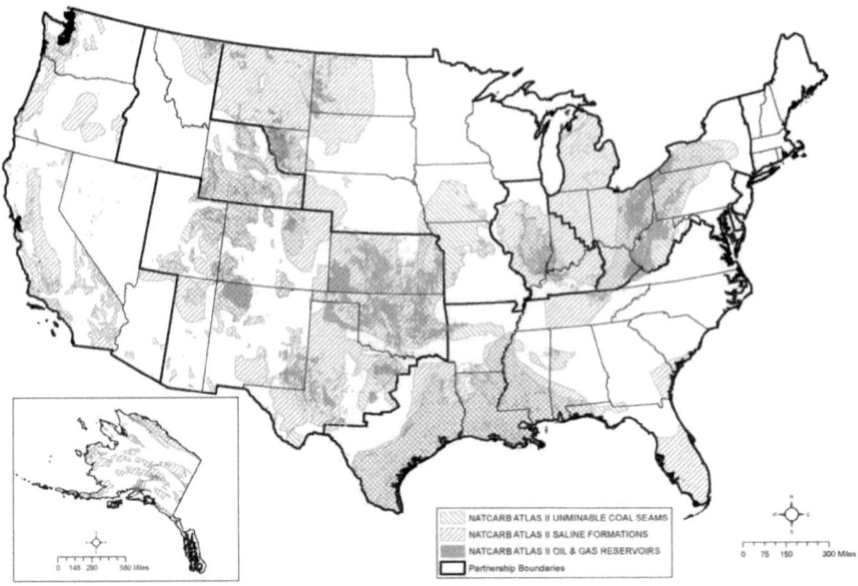

Figure 7. Potential U.S. Resources—Unmineable Coal Seams, Oil and Gas Reservoirs, and Deep Saline Formations.

Oil and gas reservoirs are well known and well defined. They have provided energy to our economy for well over 100 years. Prior to their discovery, these reservoirs retained their hydrocarbons for tens to hundreds of millions of years, an excellent demonstration of long-term storage. Some oil and gas reservoirs are still producing via the injection of CO_2 for EOR. Abandoned oil and gas fields are considered suitable for storing captured CO_2, as are active oil and gas fields in secondary or tertiary EOR operations.[20]

Unmineable coal seams are seams that are too deep or too thin to be economically mined. The definition of unmineable coal varies from region to region due to depth distribution relative to the rate and cost of mining.

Coals vary in their adsorptive properties, which are based on the physical characteristics of the coal (e.g., temperature, pressure, rank, composition). These properties affect the amount of CO_2 that can be adsorbed on the internal surface of the coal micropores. Adsorptive properties not only affect CO_2 but also other compounds within the coal. Research has shown that between 3 and

13 molecules of CO_2 can be adsorbed for every molecule of methane released, depending on the properties of the coal. Therefore, in addition to sequestering the CO_2, injection can be used for enhanced coal bed methane (ECBM) recovery by displacing methane molecules and replacing them with CO_2 molecules.[21]

The high and low values for storage potential beneath leasable Federal land for each RCSP region are posted in Table 6 and Table 7. The high estimate for storage capacity here, 374 billion metric tons, is almost 200 percent more than the low estimate of 127 billion metric tons. This is quite a variance but the distribution of potential storage space is unchanged among the RCSP between the high and low values. Big Sky region has the most storage potential followed by the PCOR region, 67 to 69 percent depending on high or low valuation. With the addition of the other western regions, 83 to 86 percent of potential storage capacity is west of the Mississippi. The SECARB region has the most storage potential east of the Mississippi. Big Sky, PCOR and SECARB are the top three regions for storage potential and retain their order between high and low estimates. The smallest region, MGSC, is also consistent here. The MRCP, WESTCARB, and SWP change order between high and low estimates. Potential storage capacity beneath Federal land is about 5 percent of the onshore storage potential for the United States.

Estimated storage potential beneath all Federal lands is 274–938 billion metric tons of CO_2 (see Table D-5 and D-6 in Appendix D). For leasable Federal lands, this represents a 50 to 60 percent reduction of potential storage from gross to leasable acreage. The vast majority of leasable acreage (98 percent, see Table 2) is controlled by BLM or the FS, which also accounts for 94 to 96 percent of the storage potential.

Deep Saline Formations

Estimated storage potential for deep saline formations beneath leasable Federal lands ranges between a low estimate of 90 billion metric tons (Tables E-3 and E-4 in Appendix E) and a high estimate of 337 billion metric tons, 71 percent and 90 percent (Table 8) of the estimated CO_2 storage potential beneath Federal land, respectively.

Table 6. Storage Potential Beneath Leasable Federal Land
(High Estimate: Million Metric Tons)

		Group 1 Total	Group 2 Total	Group 3 Total	Group 4 Total	Group 5 Total	Group 6 Total	Group 7 Total	Group 8 Total	Group 9 Total
RCSP REGION	RCSP REGION TOTALS	DOD	DOE	BLM	BOR	FWS	NPS	DOJ	Other Federal Agencies	FS and other USDA
PCOR	59,924	0	0	14,006	527	0	0	0	0	45,391
MGSC	4,459	0	0	0	0	0	0	3	0	4,456
MRCSP	9,167	0	2	0	0	0	0	0	0	9,164
WESTCARB	35,910	0	0	16,017	144	0	0	0	1	19,749
SECARB	38,803	0	12	0	1,217	0	0	4	0	37,569
SWP	34,714	0	0	19,844	195	0	0	0	3	14,672
Big Sky	191,449	0	0	170,764	20,322	0	0	0	0	363
Northeast	0	0	0	0	0	0	0	0	0	0
Total	374,425	0	15	220,630	22,405	0	0	7	4	131,365

Table 7. Storage Potential Beneath Leasable Federal Land (Low Estimate: Million Metric Tons)

RCSP REGION	RCSP REGION TOTAL	Group 1 Total	Group 2 Total	Group 3 Total	Group 4 Total	Group 5 Total	Group 6 Total	Group 7 Total	Group 8 Total	Group 9 Total
		DOD	DOE	BLM	BOR	FWS	NPS	DOJ	Other Federal Agencies	FS and other USDA
PCOR	35,325	0	0	8,841	134	0	0	0	0	26,349
MGSC	1,115	0	0	0	0	0	0	1	0	1,114
MRCSP	9,164	0	2	0	0	0	0	0	0	9,162
WESTCARB	9,066	0	0	4,056	36	0	0	0	0	4,973
SECARB	11,778	0	3	0	315	0	0	1	0	11,459
SWP	8,925	0	0	5,098	52	0	0	0	1	3,775
Big Sky	51,412	0	0	46,203	5,111	0	0	0	0	98
Northeast	0	0	0	0	0	0	0	0	0	0
Total	126,785	0	6	64,199	5,648	0	0	2	1	56,930

Table 8. Potential Storage Capacity by Reservoir Type

Reservoir Type	Range of Potential Storage Capacity (Billion Metric Tons)			
	High		Low	
Saline Reservoirs	337	90%	90	71%
Oil & Gas Reservoirs	32	9%	32	25%
Unmineable Coal Seams	6	1%	4	3%
Total	375		126	

Most of this storage potential, 60 percent to 62 percent, is found beneath BLM land with a further 31 percent to 34 percent found beneath FS land. The vast majority of storage potential beneath BLM land is found in the Big Sky region, 80 percent, followed, in order, by SWP, WESTCARB and PCOR regions. Storage potential beneath FS land is better distributed with 27 percent to 32 percent found in SECARB followed in order by PCOR, MRCSP or WESTCARB, SWP and MGSC (Tables E3 and E-4). When comparing RCSPs, Big Sky holds the largest potential for CO_2 storage in deep saline formations, a low of 53 percent and a high of 56 percent. This storage potential is mostly in the Big Horn, Wind River, and Greater Green River Basins of north central, central, and southwestern Wyoming (Figure 8). A further 43 percent to 46 percent is fairly evenly distributed between the WESTCARB, SWP, SECARB, PCOR, and MRCSP regions.

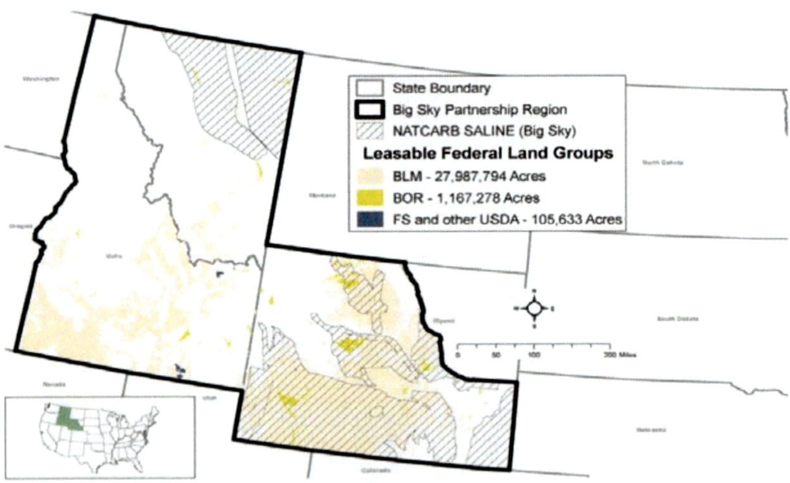

Figure 8. Deep Saline Formations and Leasable Federal Land in Big Sky.

Overall, PCOR contains the most potential here, 80 percent between BLM and FS lands. The basins of interest are the Williston and Powder River Basins of the western Dakota and northeastern Wyoming as well as some potential in north central Montana. The Big Sky region has 10 percent of the oil and gas storage potential, all beneath BLM land. East of the Mississippi, SECARB has 9 percent and MRCSP has 1 percent of the storage potential in oil and gas reservoirs, all beneath FS land. Together, PCOR and Big Sky hold 90 percent of the oil and gas storage potential beneath Federal land, about 30 percent of the potential for this type of reservoir found in the onshore Untied States.

Unmineable Coal Seams

Estimated storage potential for unmineable coal seams beneath Federal lands ranges between 6 and 4 billion metric tons, between 1 percent and 3 percent (Table 8) of the total estimated CO_2 storage potential beneath Federal land.

The vast majority of storage capacity found in unmineable coal seams, 83 to 84 percent between low and high estimates, is beneath FS lands with the remainder, 17 to 15 percent beneath BLM land (Tables G-3 and G-4, Appendix G). Most of the FS storage potential is found in the SECARB region (Figure 10), 71 percent to 77 percent with a further 21 percent to 27 percent found in the PCOR region. BLM storage potential is west of the Mississippi with 73 to 86 percent found in the PCOR regions followed by 13 to 25 percent in the Big Sky region.

Comparing RCSPs, SECARB has the most storage potential in unmineable coal seams, 59 to 65 percent, all beneath FS land. The PCOR region has 29 to 37 percent of the storage potential here split between BLM and FS land followed by the Big Sky and SWP regions, both with minor storage potential beneath BLM and FS land.

The WESTCARB region has the most Federal acreage yet considerably less storage potential than is present in the Big Sky or PCOR regions (Table 9). The distribution of Federal land and potential storage capacity (see Figure 1) in the WESTARB region has very little overlap. This is also true for the SWP region. The low percentage of Federal land present in the Big Sky and PCOR regions occurs over potential storage capacity. These two regions contain 68 percent of the saline and oil & gas reservoir storage potential beneath Federal lands. Combined with relatively low percentage of emissions, these two regions have the longest storage life, the number of years of annual emissions that can be sequestered in these regions.

Oil and Gas Reservoirs

The estimated storage potential for oil and gas reservoirs beneatł lands is approximately 32 billion metric tons, representing approxiı percent and 25 percent (Table 8) of the calculated storage potential leasable Federal lands. Oil and gas reservoirs represent production, reserves and a value for original-oil-in-place. There is considerabl knowledge of these types of reservoirs than saline or coal seam reservo interesting to note that, due to the well-characterized nature of oil a reservoirs, their storage potential is expressed by a single value. For the value of storage potential, 126 billion metric tons, this constant value ac for the much higher percent of overall storage potential beneath Federa relative to the estimate for saline and unmineable coal seams. A simil much smaller increase is also noted for unmineable coal seams.

Most of this storage potential, 70 percent, is beneath FS land wi remainder beneath BLM land. PCOR (Figure 9) has most of the FS storage potential, 85 percent (Tables F-3 and F4, Appendix F), followe MRCSP with 12 percent and the remainder in SECARB. PCOR also has of the storage potential beneath BLM land, 69 percent, with the remai found in the Big Sky region.

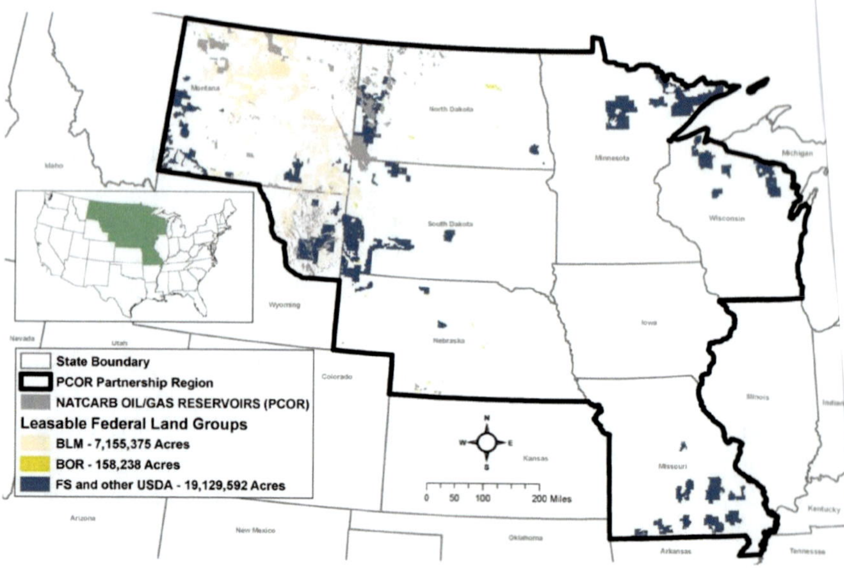

Figure 9. Oil and Gas Reservoirs and Federal Land in PCOR.

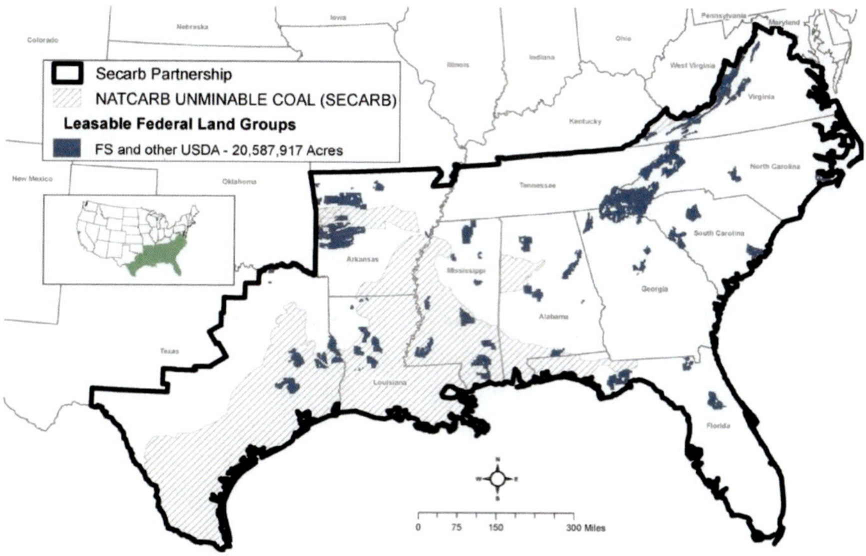

Figure 10. Unmineable Coal Seams and Federal Land in SECARB.

Table 9. Percent Leasable Federal Land, Percent Emissions On and Within 100 Miles of Leasable Federal Land, Storage Potential, and Storage Life

RCSP REGION	Percent Leasable U.S. Federal Land	Percent CO_2 Emissions On/Within 100 miles of Leasable Federal Land	Percent Total Storage Potential on Leasable Federal Land (High and Low Average)	Storage Life Potential on Leasable Federal Land (Years)
PCOR	6.6%	12.8%	21.9%	71–120
MGSC	0.4%	16.9%	1.0%	2–7
MRCSP	2.5%	20.5%	4.8%	11
WESTCARB	57.3%	5.1%	8.4%	45–177
SECARB	5.2%	27.4%	9.8%	11–36
SWP	20.2%	11.3%	8.2%	20–78
Big Sky	7.5%	4.1%	45.8%	330–1,229
Northeast	0.3%	2.1%	0%	0

Federal land occurs in large blocks in the west while eastward, in the SECAB, MGSC and MRCSP regions, Federal land occurs in smaller and more disseminated blocks. The combination of less Federal land over potential storage with much higher emissions provides for low storage life in these regions. Utilization of Federal lands here will have to be in conjunction with private lands.

Knowledge of the occurrence of potential storage capacity for CO_2 beneath Federal land as well as state and private lands is important in developing a storage field. The data presented here is the most recent submitted to NATCARB and is the basis for the second edition of the Carbon Sequestration Atlas but it only represents a potential. This database is periodically updated by the various regional partnerships. The United States Geological Survey (USGS) was directed by the Energy Independence and Security Act (EISA) of 2007 (Title VII) to conduct a carbon dioxide sequestration capacity assessment. USGS recently released their proposed procedures for conducting this assessment[22].

Potential capacity represented here will need to be further proven by actual site characterization done for the development of an actual storage field. For a developer to conduct this evaluation, they will need to have an understanding of regulations governing use of Federal lands in commercial ventures. Regulations regarding CO_2 operations are being developed at the state and Federal level. Current and pending legislation and regulations are discussed in the following section.

2.0. Laws, Regulations, and Liabilities

Numerous legal and regulatory issues concerning the storage of CO_2 beneath Federal lands must be addressed before CCS storage projects can begin. Current regulations guide oil and gas exploration and drilling, including injection of water and CO_2 for secondary and tertiary EOR recovery, as well as injection of natural gas for storage. In addition, a few states and the U.S. Environmental Protection Agency (EPA) have taken preliminary actions to address the unique challenges presented by permanent CO_2 sequestration. This study reviewed existing and proposed laws and regulations that may support and/or impact CCS operations. The areas examined include environmental considerations, Federal land leasing, pipeline rights-of-way, pipeline operations and safety, CO_2 injection and storage, and short- and long-term liability.

2.1. Federal Land Leasing

Large tracts of Federal lands, by controlling both surface and subsurface rights, can convey with a lease of sufficient acreage a corresponding subsurface area representing a potentially large reservoir that may be well suited for large-scale storage projects. Negotiating with several or even numerous landowners, depending on the size of the proposed storage field, will be a time-consuming effort on the part of the project developer. Acquiring a lease from a single landowner may significantly reduce the amount of time required to secure the necessary surface and subsurface rights for a storage project.

Approximately 56 percent of the land owned by the United States and managed by various Federal agencies is leasable for use by private interests. Two Federal groups, the Bureau of Land Management and the U.S. Forest Service, manage 98 percent of the leasable acreage available. The Department of Energy and Bureau of Reclamation manage the remaining 2 percent.

In most cases, BLM acts as the leasing agent for mineral rights beneath both BLM and FS lands (on-shore). The Forest Service determines where and under what conditions oil and gas leasing can occur on National Forest System lands. BLM then ensures that NEPA requirements have been met before it offers its own and Forest Service oil and gas leases at auction.

BLM conducts Federal land leasing during quarterly oil and gas lease sales. A BLM form is completed by an applicant for public domain lands or acquired lands and includes a legal description of the land requested by a parcel number identifier. A competitive oral auction is held, and the Federal land lease is awarded to the highest bidder. Federal land not bid on at auction becomes available for noncompetitive bidding.

Leases for BLM-managed land are pursuant to the Mineral Lands Leasing Act of 1920, as amended and supplemented[23] and the Mineral Leasing Act for Acquired Lands of 1947, as amended[24], as detailed on the "Offer to Lease and Lease for Oil and Gas," Form 3100-11[25]. As discussed above, acreage available from BLM or the FS may carry further restrictions to access due to wildlife and/or ecosystem considerations.

National Forest Plans and BLM Resource Management Plans contain many of the stipulations that are included in leases. Some Federal lands are unavailable for leasing. Areas of Critical Environmental Concern, such as wilderness areas and national parks and monuments, are unleasable, and national forest inventoried roadless areas require regulations to render them accessible.

An Area of Critical Environmental Concern is a location that receives special management attention because of potential hazards and/or to protect and prevent irreparable damage to important historic, cultural, or scenic values, fish and wildlife resources, or other natural systems or processes.[26] Wilderness areas, as well as national wild and scenic rivers, are Congressionally designated areas protected for the preservation of their natural condition. A BLM wilderness area is public land that Congress has designated for BLM to manage as a component of the National Wilderness Preservation System in accordance with the Wilderness Act of 1964.[27] No surface or subsurface lease for CCS would be attainable for any federally owned wilderness areas in the United States.

The unavailability of surface leases is also true for lands designated as national parks or monuments. However, "split estates" exist in which the Federal government owns the surface rights, but mineral rights are privately owned. In this case, the Federal government cannot deny access to privately held minerals, but can regulate mineral activities to a varying degree. These regulations are site specific.[28]

FS lands include national forest inventoried roadless areas. These areas present barriers to private development; however the Roadless Area Conservation Rule (May 2005) allows governors to petition the Secretary of Agriculture to develop regulations to manage roadless areas to meet state-specific needs. Leases have been issued in various forests within inventoried roadless areas, although the disposition of roadless areas is unresolved and potential exists that no roads may be allowed by future decision(s).[29]

2.2. Environmental Considerations

Laws such as the National Environmental Policy Act (NEPA),[30] the National Historic Preservation Act (NHPA),[31] and the Endangered Species Act (ESA)[32] apply to Federal actions that authorize private activities on public lands, such as the issuance of leases and pipeline ROWs. Some require reclamation. Sequestration projects on Federal lands will likely be subject to these types of stipulations. They will also be subject to a yet-to-be-established application process and its associated impacts to project costs and schedule.

NEPA requires Federal agencies to study the environmental effect of their actions through an interdisciplinary planning process. For situations with potentially significant environmental effects, the NEPA process informs and seeks input from the public, tribes, states, and local agencies, as well as other Federal agencies.[33]

NHPA contains responsibilities for Federal agencies to establish preservation programs and designate Federal Preservation Officers for the coordination of historic preservation activities. NHPA requires archaeological surveys to reveal the presence of historic sites that require protection or relocation.

Congress passed ESA in 1973 to conserve the ecosystems that sustain endangered and threatened species. Congress considered such fish, wildlife, and plant species to be "of esthetic, ecological, educational, historical, recreational, and scientific value" to the Nation.[34]

Drilling Permits and Conditions of Approval

There are just over 152,000 oil and gas wells of various vintage located on Federally managed land, each of which falls under a lease agreement between the drilling agency and the managing agency. Each lease and drilling permit contains conditions of approval (COA) to achieve environmental protection, all of which are subject to change at any time during the extent of the lease. These COAs reflect the stipulations associated with the lease upon which a drilling permit is sought. Depending on site attributes, these stipulations and COAs are subject to change at any time during the extent of the lease.

Plans or COAs for specific geographic locations include factors that could affect the injection of CO_2 for sequestration. Drilling may be limited to a few months or specific seasons per year in consideration of a threatened, endangered, or sensitive species. A historic site may need protection or relocation. Air and water quality may need to be addressed. Or proposed activities may raise concerns about the sights and sounds of a scenic area. Such situations often require stipulations that further restrict access or reduce an amount of acreage available for lease.

In 2003, the National Petroleum Council (NPC) conducted a study regarding environmental, wildlife, cultural, and other surveys, all components of COA's, necessary to securing oil and gas drilling permits. The NPC found that such surveys add an additional $15,000 to $250,000 and 1 to14 months to initiating field operations for an exploration well. Survey costs for oil and gas development wells were lower, in the range of $17,000 to $108,000 each, with a concurrent time delay range of 1 to 22 months.[35]

NEPA is an important tool in the Federal land management decision-making process, encouraging productive and enjoyable harmony between man and the environment. These impacts may be evaluated by one of four types of review: categorical exclusion, documentation of NEPA adequacy, environmental assessment (EA), and environmental impact statement (EIS).

Impact evaluations are likely to be performed for CCS projects via EAs and EISs. For CCS activities, these evaluations would include the effects of both accessing a drilling location (e.g., pipelines and roadways) and utilizing the actual injection site. Completion of an EA and EIS require detailed input on project specifics; for example, a geologist or engineer may need to supply the necessary inputs. EA and EIS evaluations may also require a significant financial commitment by the operator and involve extensive periods of time. Where the development and approval of a simple EIS may span only two years, one with complex issues could take up to six years for completion.

A range of statutory and discretionary requirements beyond those specified as lease stipulations have potential to impact CCS drilling. These requirements may be derived from evaluations required by NEPA. As mentioned earlier, an EIS can delay the implementation time of the drilling operation approximately 2–6 years, depending on the complexity of associated issues. Along with the performance of the EIS, the land-use planning process will also require additional time.

ESA designates three categories of protected species: listed, proposed, and candidate. Federal agencies are required by the ESA to conserve listed species. BLM and FS jurisdictions treat listed and proposed species in a similar manner. In addition, Federal agencies manage wildlife habitats, and not all habitat areas have been mapped. CCS drilling permits could be delayed by incomplete habitat mapping if this information is required before leasing. Conservation of listed species is achieved in critical habitats, which may require special management considerations or protection. Federal agencies ensure that projects are permitted in a manner unlikely to result in the destruction or adverse modification of critical habitat. This is expressed in the Endangered Species Act: "Critical habitat may be established for species now listed as threatened or endangered for which no critical habitat has been established. Except in circumstances determined by the Secretary, critical habitat shall not include the entire geographical area which can be occupied by the threatened or endangered species."[36]

Seasonal restrictions might also exist for drilling activities for the protection of wildlife. For example, in Alaska an approximate time period of 5 months is available in the winter for exploratory drilling activities. This time period can be further shortened by activities such as ice road building. Exceptions could be granted on a case-by-case basis if conditions would change. Other examples of seasonal restrictions are:

- San Juan River Basin Plan, New Mexico. No development activities during spring calving and raptor roosting.
- Jack Morrow Hills Plan, Wyoming. Protections for crucial winter and parturition areas for elk and mule deer. Seasonal restrictions for raptor nest sites and winter roosting, as well as "no surface occupancy" for raptor nesting areas.
- Otero Mesa Plan, New Mexico. Restrictions and controlled surface occupancy for raptor nest sites.

A "no action" alternative is also required by NEPA. "No action" is an evaluation performed to determine the consequences of *not* performing the proposed project. The intent of such an evaluation is to provide a reference against which decision makers may gauge the value and merit of implementing the proposed project. The "no action" alternative maintains established trends or management direction and implements those actions previously analyzed and/or approved. This type of alternative must be evaluated in addition to the performance of analyses for other alternative actions for NEPA documents. Direct, indirect, and cumulative impacts (positive and negative) on resources (e.g., ground and surface waters, flora and fauna, other recoverable resources) as well as to counties and states also need to be considered for CCS projects.

Suburban Encroachment

A final issue that could affect CO_2 CCS field development is categorized by BLM and the FS as "suburban encroachment." The issue involves no surface occupancy by drilling apparatus near residential housing developments where opposition is likely. Future CCS projects will need to include this emerging issue in their plans.

2.3. Pipeline Rights-of-Way

Establishment of a ROW on Federal land is a pipeline regulatory issue and as been described as:

> "A permit or easement which authorizes the use of public land for certain specified purposes, commonly for pipelines, roads, telephone lines, etc.; also, the lands covered by such an easement or permit. It does

not grant an estate of any kind, only the right of use. May also include a site."[37]

The objective of BLM's ROW program is to grant ROWs to individuals, businesses, or government entities and to direct and control their use. A pipeline ROW through any Federal land may be granted for the transportation of oil, natural gas, synthetic liquid or gaseous fuels, or any refined product.[38] The ROW can include contiguous facilities, such as valves, pump stations, supporting structures, bridges, monitoring and communication devices, surge and storage tanks, terminals, roads, airstrips, and campsites. However, when these facilities are not connected or contiguous to the pipe, they may be the subjects of separate ROWs.[39]

The length of time for a ROW is requested on the pipeline ROW application and is dependent on the specific project length, volume capacity, or even if the project is temporary. The BLM considers several factors in determining a reasonable term. These include the public purpose served; cost and useful life of the facility; time limitations imposed by licenses or permits required by other Federal agencies and state, tribal, or local government; and the time necessary to accomplish the purpose of the grant. For the purpose of CCS, the life of the ROW needs to be greater than the project life. Following is a discussion of some challenges in obtaining a pipeline ROW for CCS projects.

Environmental Impact Statements and Pipeline Rights-of-Way

Building pipelines through Federal land requires an EIS, which typically includes an exhaustive public participation component, such as public hearings and comment periods. All Federal lands are open for ROW access, but ROWs across otherwise unleasable acreage must meet strict stipulations. For example, surface leases are unavailable in national parks, but Congress has granted ROWs in these areas provided that they do not interfere with scenic corridors.[40]

NEPA,[41] NHPA,[42] and ESA[43] affect whether a ROW is granted, and they may require that specific conditions or limitations be included in the grant of any particular ROW.[44] To reduce the time required for individual projects to obtain approval for ROWs on Federal land, a draft programmatic EIS (PEIS) was performed for the designation of Federal energy corridors in 11 western states. The draft PEIS analyzed the environmental impacts of designating corridors for oil, gas, and hydrogen pipelines and electricity transmission and distribution facilities (energy corridors). The energy corridor designations were also incorporated into relevant land use and resource management plans.[45]

Grant Authorization

The Federal Lands Policy and Management Act (FLPMA) of 1976 [46] authorizes the Secretary of the Interior (public lands) and the Secretary of Agriculture (National Forest System lands) to grant, issue, or renew ROWs that include pipelines. Under the Act, terms and conditions are included in each grant. Part of the purpose of terms and conditions is to "minimize damage to scenic and esthetic values and fish and wildlife habitat and otherwise protect the environment."[47] Other terms and conditions may be imposed to protect Federal property and economic interests.[48]

The FLPMA contains other provisions for the Secretaries to promulgate regulations, require advance rental payments, and impose bonding requirements, among other duties.[49]

Congress has also provided executive branch agencies with authority to grant ROWs on Federal lands that are within their control but not covered by FLPMA.[50] Public Law No. 87-852 was recodified by Public Law No. 107-217 to provide the Federal government with an economical and efficient system to use available property.[51]

Eminent Domain

Acquiring the rights to surface access from the owner of the desired property is best accomplished by voluntary negotiations. If this is not successful, other means are available. If applicable within the state, the right of eminent domain authorizes the State to seize private property without the owner's consent. For pipeline ROWs this power can be delegated to a private utility for the construction and operation of pipelines.

At the present time, Federal eminent domain authority does not extend to CO_2 pipelines. Several states have provided eminent domain authority to CO_2 pipeline owners to assist in getting CO_2 pipelines constructed.[52] For example, common carrier pipelines in Texas have a statutory right of eminent domain. Common carrier pipelines are operators that transport oil, oil products, gas, CO_2, salt brine, sand, clay, liquefied minerals, or other mineral solutions.[53]

Temporary Permits

A ROW may be supplemented by temporary permits for the use of Federal lands in the vicinity of the pipeline to provide for its construction, operation, maintenance, or termination, or to protect the natural environment or public safety.[54] ROWs or permits granted or renewed are subject to stated regulations and to any terms or conditions regarding extent, duration, survey, location, construction, operation, maintenance, use, and termination.[55]

For example, temporary use permits (TUPs) are necessary for extra construction width that may be needed during construction of a pipeline. The additional land may also be necessary for stockpiling of excess materials and equipment parking. The TUPs are applied for by the pipeline constructor (or other applicant representative of the CO_2 pipeline) at the time of application for the pipeline ROW via standard form 299 (SF-299) with BLM. The TUP may also be applied for on a separate SF-299 form after the ROW is granted. An additional processing/monitoring fee would apply.[56]

Right-of-Way Corridors

To minimize adverse environmental impacts by ROWs and prevent the proliferation of separate ROWs across Federal lands, the need for a national system of transportation and utility corridors across Federal land was discussed in the 1970s.[57] In 1976, Congress addressed the designation of ROWs in Section 503 of the FLPMA. The Western Utility Group has worked with BLM and the FS since 1979 to further delineate ROW corridors, but with limited success.

In 2001, President George W. Bush issued Executive Order Number 13212 to create the White House Task Force on Energy Project Streamlining. Additional efforts were pursued by BLM, the Forest Service, the Western Governor's Association, and various utility groups. More work is needed in Federal planning areas to complete the task.

Section 368 of The Energy Policy Act of 2005 (EPACT05) required the Secretaries of Energy, Agriculture, Interior, Commerce, and Defense to designate corridors for oil, gas, and hydrogen pipelines and electricity transmission facilities on Federal land and in 11 contiguous western states[58] (see Figure 11). The timeliness of acquiring individual ROWs has been improved by this requirement by precluding the need to consider alternative routes.

Section 368 also requires the Agencies to conduct any environmental reviews necessary to complete the designation of Section 368 energy corridors. The proposed designation of Section 368 energy corridors would not result in any direct impacts on the ground that may significantly affect the quality of the human environment.[59] Although EPACT05 Section 368 duplicates Section 503 of the earlier-enacted FLPMA, Section 368 provides more specificity regarding the ROW and imposes deadlines. Under Section

368 of EPACT05, the designation of energy corridors for western states was due not later than two years and for other states not later than four years after enactment of EPACT05. Since November 2007, public hearings have been held in all 11 of the western states and in Washington D.C. for the West-wide Energy Corridor Draft Programmatic EIS.

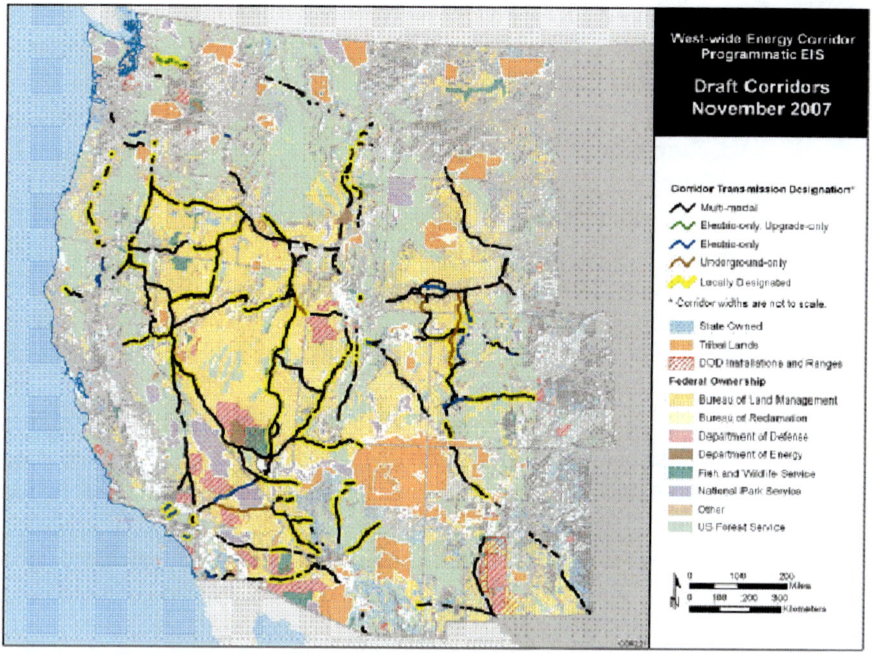

Figure 11. Eleven-State (Western) Map of Proposed EPACT05 Section 368 Energy Corridors.

This collaborative effort prepared a draft PEIS pursuant to NEPA that identifies the impacts of designating energy corridors that could contain various pipelines and electricity transmission facilities. The officially designated energy corridors require compliance with applicable laws and regulations. An important factor in the energy corridors was the use of existing corridors and/or ROWs to minimize new corridor locations.

State Standards and Incorporation of Rights-of-Way Requirements from Various Sources

To the extent practical, the Federal agency granting a ROW is required to comply with State standards for ROW construction, operation, and maintenance.[60] For example, Alaska's Commissioner of the Department of Natural Resources analyzes ROW applications, negotiates/drafts ROW lease documents that include mandatory covenants, and determines whether to offer a specific lease to a specific applicant. In making a determination, the Commissioner takes into consideration such factors as environmental impacts and protecting the interests of individuals who live in the general area of the ROW and rely on fish, wildlife, and biotic resources of the area for subsistence purposes.[61] Federal agencies seeking to grant ROWs are subject to the same process followed by the Commissioner.

Some states grant power to municipalities to regulate the construction of public ROWs. Examples of these include:

- Alabama, which maintains the power of municipalities to regulate construction of public ROWs and write ordinances accordingly
- Alaska, which grants municipalities the power to regulate ROWs
- Ohio, which grants authority to municipalities to manage ROWs, recover fees, and promote municipal ROW coordination and standardization
- Pennsylvania, wherein municipalities may grant licenses to public service companies to construct lines if those lines will give state buildings better service, or if such lines are necessary to serve the public
- Wyoming, in which local governments are granted authority to take all necessary actions to plan, construct, maintain, and regulate the use of streets, including the regulation of any structures thereunder[62]

It is also common practice at the State level to develop requirements and/or policy by reference. That is, instead of replicating Federal or other standards that may change in the future, states frequently rely on the desired standard by "referencing" it in the policy or regulation.

Referencing eliminates the need for states to amend regulations should the referenced regulation change. For example, to accommodate utility facilities crossing highways under the jurisdiction of the Wyoming Department of Transportation, basic and operating policies were developed by incorporating Federal laws, guidelines, rules, regulations, and industry standards "by reference," as well as State and industry standards.

2.4. Pipeline Operation and Safety

The U.S. Natural Gas Pipeline Safety Act allows the Federal government to regulate interstate transportation and storage of natural gas.[63] State governments are allowed to regulate intrastate pipelines, including locations, as long as their regulations meet and do not conflict with minimum Federal regulations.

The Secretary of Transportation has primary authority to regulate interstate CO_2 pipeline safety under the Hazardous Liquid Pipeline Act of 1979 as amended[64]. Because CO_2 is transported at high pressure, lower-pressure natural gas pipelines are unsuitable for CO_2 transport. The U.S. Department of Transportation (DOT) regulates the design, construction, operation and maintenance, and spill-response planning for CO_2 pipelines[65] and administers pipeline regulations through the Office of Pipeline Safety within the Pipelines and Hazardous Materials Safety Administration.

Although CO_2 is listed as a Class 2.2 (non-flammable gas) hazardous material under DOT regulations, the agency applies nearly the same safety requirements to CO_2 pipelines as it does to pipelines carrying such hazardous liquids as crude oil, gasoline, and anhydrous ammonia.[66]

2.5. Injection of CO_2

For long-term storage, the injection of CO_2 into the subsurface is not specifically addressed by Federal regulations, although the EPA recently released proposed rules for Class VI wells, a new classification for the Underground Injection Control (UIC) program. The UIC program operates under the Federal Safe Drinking Water Act (SDWA)[67] and provides regulations for CO_2 injection for EOR projects.

The primary purpose of UIC rules is to protect U.S. drinking water (USDW) or groundwater with less than 10,000 parts per thousand total dissolved solids. The EPA currently operates five classes of injection wells based on similarity in fluids injected, activities, construction, injection depth, design, and operating techniques. This categorization ensures that similar wells meet appropriate performance criteria for protecting USDW.[68]

Prior to passage of the SDWA of 1974, State regulations governing injection wells provided a disparate level of protection to the ground waters that provided drinking water to the public. The regulations written by EPA to implement the statutory requirements of the SDWA establish minimum requirements for controlling all injection activities and provide mechanisms for the implementation and authorization of primary enforcement authority:

> "State and Tribal governments can apply for primary enforcement authority (primacy) for the UIC Program under the procedures provided in Federal regulations (Title 40 Code of Federal Regulations (CFR) Part 145). For a state to assume primacy and receive implementation funding, the state must demonstrate (Section 1422 of the SDWA) that it has authority over injection activities on Federal lands (except for Indian Country). Should states or Tribes decide not to seek primacy, EPA implements the program (See House Report No. 93-1185, July 10, 1974 and Section 1421 of the SDWA)."[69]

Out of the five UIC well classifications, the following currently regulate all present-day CO_2 injection activities throughout the United States:

Class I—Injection of hazardous wastes, industrial non-hazardous liquids, or municipal wastewater beneath the lowermost USDW. Because CO_2 is not currently regulated as a hazardous material, this class may be used for CO_2 injection as a non-hazardous waste. RCSP wells are designed to this class level even though the permit issued may not be for a Class I well. Modeling is required to show that the injected fluids will stay in place for a period of 10,000 years.

Class II—Injection of natural gas for storage and brines and other fluids associated with oil and gas production. Injection occurs beneath the lowermost USDW. This classification encompasses CO_2 injection for EOR operations. Existing oil and gas wells present conversion possibilities for the injection of CO_2 on Federal land, or new CO_2 injection wells could be drilled. Wells drilled to a storage reservoir also present possible breaches in the cap rock, providing potential migration pathways into overlying horizons or possibly to the surface.

Class V—All current injection wells not included in Classes I–IV. In general, Class V wells inject non-hazardous fluids into or above USDWs and are typically shallow, on-site disposal systems. However, some deep Class V wells inject below USDWs. This class would only encompass experimental CO_2 injection or injection of CO_2 above the lower-most USDW not related to oil and gas production. The U.S. EPA issued interim guidance, pending a final rule, that injection wells used for demonstration and research of CO_2 sequestration should be permitted under this UIC class.[70]

Class VI (proposed)—The EPA recently released its proposed rules for Class VI wells for injection of CO_2 for sequestration. They are published in the Federal Register.[71] Although not yet finalized, these rules provide an idea of an operator's injection and post-injection responsibilities relative to the area of review. Areas of responsibility will likely include initial characterization of a proposed field, monitoring, verification, and accounting (MVA) of the CO_2 injection and plume, and plugging injection wells; the proposed rules also cover the period of time the operator would be required to monitor the CO_2 plume after injection is complete.

In seeking input for Class VI injection rules, the EPA pointed out that other constituents in the CO_2 stream may trigger hazardous material clauses under Resource Conservation and Recovery Act (RCRA). Also, the Comprehensive Environmental Response, Compensation, and Liability Act (CERCLA), may come into play should the CO_2 plume migrate beyond its designated boundaries. EPA noted that the composition of the injected CO_2, as well as the estimated boundaries of the CO_2 plume upon cessation of injection, would have to be carefully considered when filing for a permit.

Once injection is complete and the well is to be plugged and abandoned, the proposed rules recommend a period of 50 years over which to establish non-endangerment before the operator is released from liabilities related to the well. This period could be shorter or longer depending on when "non-endangerment" is established. This post-injection period of time is important in addressing long-term liability issues. The EPA intends to issue final rules in 2011.

2.6. Storage

Storage of CO_2 within a reservoir in the subsurface requires the operator to have control of the pore space, to have the right to inject CO_2 into this pore space and leave it there. A storage field operator will need this right of ownership in order to conduct injection operations in the short-term. Once injection of CO_2 is finished, then the storage field will begin a transition to longterm storage. This study reviewed existing and proposed Federal and State laws and regulations for their applicability to potential storage of CO_2 beneath Federal lands, both in terms of short-term (injection phase) and long-term (post-injection phase) periods, as well as situations involving trespass.

Long-term storage has unique challenges related to the anticipated scale of the projects. The amount of storage space (volume) involved will be large, and the post-injection time period could have a range of hundreds to thousands of years. Long time periods for CCS will most likely require a transition of liability from a storage operator to another entity. The particular nature of this entity has yet to be determined, but most believe that government, at either the State or Federal level, will be involved.

Regulations providing guidance—at either the Federal or State level, for either private or public operations—have yet to be provided. What is generally agreed is that the operator should be relieved of obligations and liabilities and the storage field decommissioned at some point in time after injection operations cease. This will occur when "non-endangerment" can be established.

Pore Space Ownership

Two property interests are present in determining a right to access and produce hydrocarbons, oil, natural gas, or coal, stored in the subsurface: mineral interest and surface interest. The rights conveyed here are for extraction. CO_2 sequestration, however, involves injection for permanent storage; the intent is that injected CO_2 will occupy subsurface pore space for ever.

Pore space and subsurface formations belong to the surface owner in a majority of states. This is known as "fee simple." Except in the case of split estate, the government is the owner in fee on Federal lands, meaning that everything—surface, minerals, subsurface, and pore space—is owned by the United States. If the pore space beneath Federal land is leased, control of the pore space belongs to the lessee until the expiration of the lease period.

Expiration of a lease for injection of captured CO_2 for sequestration will have very different implications from that encountered in oil and gas operations. Ownership conveys rights as well as responsibilities and liabilities. Liabilities associated with CO_2 storage operations and post-injection long-term storage has yet to be sorted out. Ownership of the injected CO_2 has yet to be determined. Presently, Federal leases are designed for mineral extraction, not for long-term storage of captured CO_2. Proposed Federal legislation addresses several aspects of CO_2 sequestration but nothing specific has been enacted.

As mentioned earlier, Wyoming recently passed legislation regarding CO_2 storage that addresses pore space ownership (JIB No. 0089). Wyoming is the only state thus far to address this ownership issue. The term "pore space" was defined to mean subsurface space that can be used as storage space for CO_2 or other substances. It declared the ownership of all pore space in all strata below the surface lands and waters of the state to be vested in the several owners of the surface above the strata. The pore space would accompany conveyance of the surface ownership of real property unless the ownership interest in such pore space was previously severed from the surface ownership or is explicitly excluded in the conveyance. Further, ownership of pore space in strata may be conveyed in the manner provided by law for the transfer of mineral interests in real property.

In Wyoming, conveyance of any mineral or other subsurface interests does not convey ownership of any pore space in the stratum unless the agreement explicitly conveys that ownership interest. The owner of pore space does not have any right to use the surface estate unless it is addressed in the recorded instrument or unless the pore space owner is also the surface estate owner.

After July 1, 2008, transfer of pore space rights are null and void at the option of the owner of the surface estate if the transfer instrument does not contain a specific description of the location of the pore space being transferred. Pore space rights do not affect the respective liabilities of any party. Subsurface pore space rights acquired prior to July 1, 2008, remain valid after the effective date of this bill.

Saline Formation Pore Space Ownership

Saline formations are expected to be used widely for CCS projects due to their widespread availability and potential for storage. This type of storage reservoir provides the greatest potential for sequestering captured CO_2 beneath both Federal and non-Federal lands. Mark de Figueiredo's report on property interests and geologic CO_2 storage provides property rights insight relative to ownership of saline formations.

> ". . . there is an inherent uncertainty concerning the determination of property rights for a saline formation with respect to CO_2 storage because of the lack of case law on point. Instead, the law has focused on property rights over the taking and use of groundwater for consumption. . . . The determination of property rights over a saline formation is comparable to the mineral formation case. In the majority of states, the owner of the surface has the right to make any use of the subsurface space, including the saline formation. Just as in the case of a mineral formation, where ownership of non-depleted minerals must be accounted for, any storage operation needs to take into account ownership of the water contained in the saline formation. There are a number of property regimes that states use to determine property rights over the water. In general, states follow one of five major doctrines: absolute dominion, reasonable use, prior appropriation, correlative rights, or the Restatement rule."[72]

2.7. Liability

Injection of CO_2 on Federal land will likely involve many responsibilities for the operator and be divided into short-term and long-term liabilities. Liabilities during injection in the short term are expected to be similar to those related to oil and gas operations. The important difference is that oil and gas operations leave behind a pressure-depleted reservoir, whereas injection of captured CO_2 for long-term storage will leave a reservoir at pressure.

Liabilities in the long term present a challenge because the permanent sequestration of captured CO_2 in the subsurface is a new concept. Injection operations can occur across many decades, depending on the size of the project. Post-injection monitoring may last five decades. After so many years, who will be in a position to assume liability for the long-term assurance that the stored plume of CO_2 in the subsurface remains harmless? This is very different from oil and gas operations, although orphaned well programs at the state level can provide an analogy. Trespass is a liability that can occur during operations or post-operations is migration of the CO_2 plume into areas not considered or tracked by MVA techniques. The following sections provide a list of potential hazards that can become a liability, including trespass.

Short-Term Liabilities

Short-term liabilities render the lessee responsible for events that occur during siting, construction, operation, closure, and post-closure (term of lease timeframe). Examples of short-term liabilities include:

• Injury to workers	• Public exposure to CO_2
• Permit acquisition	• Lawsuits
• Acquisition of access or storage rights	• Improper MVA installation
• Cost overruns	• Materials failure
• Contractor delays	• Induced seismicity
• Improper well abandonment	• Atmospheric release
• Poor well construction	• Groundwater confinement zone failure
• Drilling wells into areas of low injectivity	• Groundwater affected by CO_2 interaction with subsurface elements
• Failure to adequately complete old wells/boreholes	• Groundwater affected by displaced brine

• Surface property damage (Federal or private)	• Damage to a confinement zone (for example by fracturing)
• Subsurface property damage (mineral rights)	• Environmental damage resulting from a surface pipeline leak and/or well blowout

Less predictable liability components of a CCS project may include cost increases due to inflation, resource constraints, and changes in regulations. Short-term risk management is essential in determining a project's viability for the investor(s)/financier(s). Without a manageable level of cost certainty, project economics will not be favorable and the project likely not funded.

All of these short-term hazards will have to be identified. Reducing their risk of occurrence will limit the associated liabilities and satisfy the concerns of an insurance company. A prospective operator will have to prove financial responsibility prior to gaining a permit or certification of the project to begin active injection operations. This effort establishes a plan for safe operation of injection activities. Maintenance of this safety plan throughout operations should also help mitigate long-term liabilities.

Long-Term Liabilities

Long-term liabilities begin post-injection, after cessation of storage field operations. Injection pressure will no longer be applied to the storage formation. Liabilities here arise from migration of the CO_2 plume, either vertically through well bores, fractures, or faults or horizontally by moving to similar points of leakage. Adverse impacts may include:[73]

• Public exposure to CO2	• Change in law/regulation
• Subsurface property damage (mineral rights), trespass	• Third-party damage to confinem zone
• Ecosystem degradation (terrestrial or aquatic)	• Atmospheric release (loss of credits/compliance)
• Groundwater contamination by displaced brine or gas	• Lawsuit

The ability of storage field operators to gain a release from the liability associated with their operations depends on how well they reduce the risks associated with the hazards of operations. Knowledge of the CO_2 plume in the reservoir is based on years of MVA operations. After decades of injection operations, a well-characterized CO_2 plume and its associated hazards should be well recognized. With establishment of non-endangerment, the long-term liability should be acceptable by another entity. The need for successful storage and MVA operations on Federal lands is as important as that for private lands.

Trespass

Two types of geophysical subsurface trespass and associated liabilities with CO_2 may occur in the short or long term. Subsurface trespass can occur when someone produces or otherwise drains the stored CO_2 from the storage reservoir. Subsurface trespass may also be caused by migration of CO_2 into an area beneath a neighboring property where access and associated rights were not leased. Subsurface trespass liability would depend on whether the trespasser acted in good faith when production or drainage of CO_2 occurred in the subsurface.

The intrusion of CO_2 onto a neighboring property may cause damage and trigger compensation if the adjacent owner can prove he or she has suffered damages. In Chance v. BP Chemicals and Mongrue v. Monsanto, the courts recognized subsurface trespass of substances injected through wells and allowed subsurface trespass as a cause of action[74]. The plaintiffs in these cases, however, were unable to show that intended use of the subsurface property was compromised and that damage occurred. Neither plaintiff was able to meet their burden of proof, so neither was awarded compensation. CCS project operators must be mindful that subsurface trespass could occur through the migration of CO_2 into areas inside or outside the reservoir that initial modeling did not anticipate.

The oil industry has addressed this liability during EOR and the issue of trespass has been addressed in a Texas case (Texas Railroad Commission v. Manziel)[75], which held that injection associated with a state-authorized secondary recovery project would not cause trespass. This was decided even though fluids move across property lines.

Under the Mineral Leasing Act of 1920, an independent contractor is liable to the United States for any losses caused by any intentional or reckless action or inaction. The independent contractor is required to maintain a bond commensurate with the amount of money for which such individual could be liable to the United States.

2.8. Legislative Activity

At the Federal and state levels, some legislation regarding storage of capture carbon dioxide has been enacted while other proposed bills have either reached the floor for debate or failed to make it out of committee. Legislation addressing liabilities associated with storage operations are proposed at the state and Federal level. Only Texas and Illinois have address this issue but only with respect to proposed FutureGen projects proposed in each state. Some other storage-related regulations and laws are presented at the end of the section.

Enacted and Proposed Federal Legislation

There is considerable activity at the Federal level for proposed legislation and regulations. The EPA's proposed rules for Class VI injection wells, once finalized and adopted, will provided guidance and minimum standards across all states at the Federal level. Several proposed pieces of legislation in Congress would direct the EPA to protect USDW under the Clean Water Act. EPA's efforts here anticipate these concerns, which were discussed earlier.

The recently enacted Energy Independence and Security Act of 2007 (P.L. 110-140) provides for large-scale demonstrations of CCS technology, encouraging a preference toward the combination of industrial sources with transportation and sequestration in a single demonstration, if possible.[76] This bill directs DOI to conduct a national assessment of onshore CO_2 storage potential and, as mentioned earlier, USGS recently released their report on *Development of a Probabilistic Assessment Methodology for Evaluation of Carbon Dioxide Storage*. DOI is also directed to submit a report on a recommended regulatory framework for managing geologic carbon sequestration on public lands including, but not limited to, ensuring fair market values, public participation, protecting natural and cultural resources, status of liability issues, legal and regulatory issues, split estates, pipeline ROWs, and leasing.[77]

The proposed Lieberman-Warner Climate Security Act of 2008[78] (S.3036, formerly known as S.2191) would have required an assessment of geologic storage potential. This bill would have also required an assessment of and the feasibility of constructing pipeline and sequestration facilities. Of particular importance, this bill would establish a task force to evaluate the "potential Federal development of a framework for geological sequestration of CO_2."

The proposed Carbon Capture and Storage Technology Act of 2007[79] (S.2323), although designed primarily to provide for CCS research, development, and demonstration projects, also addresses CCS regulations and the establishment of an Interagency Task Force to develop regulations providing guidelines and practices for CCS. This bill would also create a Task Force composed of the Secretary of Energy, the Administrator of the EPA, and the Secretary of the Interior (acting through the Director of the USGS). Consultation with industry experts, legal experts, and technical experts is specified. Requirements for development of the regulations by the Task Force are listed and are to take into account existing EPA-UIC program requirements. Other requirements to be addressed by the Task Force are:

- Certification and closure of CCS sites; potential appropriate transfer of liability to governmental entities
- Mechanisms to ensure, monitor, and verify the safe transportation and storage of CO_2
- Estimation of the costs of carrying out the regulations
- Taking into account the outcomes of demonstration projects

In this bill, the Task Force has three years to complete their assigned work. Following submittal to Congress, the EPA is charged with the responsibility of promulgating these rules within 18 months.

The proposed Carbon Dioxide Pipeline Study Act of 2007[80] (S.2144) would require DOE, the Federal Energy Regulatory Commission, DOT, EPA, and DOI to assess the feasibility of the construction and operation of pipelines and CO_2 sequestration facilities. These pipelines would be used for either sequestration or EOR. The study would consider the following:

- Barriers or potential barriers in existence, such as technical, siting, financing, or regulatory barriers relating to the construction and operation of the CO_2 pipelines and CO_2 sequestration facilities
- Market risk (including throughput risk) relating to the construction and operation of the CO_2 pipelines and CO_2 sequestration facilities
- Regulatory, financing, or siting options that, as determined by the Secretary of Energy, would mitigate market risk or help ensure the construction of pipelines dedicated to the transportation of CO_2
- Means by which to ensure the safe handling, transportation, and sequestration of CO_2
- Preventive measures to ensure the integrity of CO_2 pipelines
- Other appropriate issues as determined by the Secretary of Energy

Enacted and Proposed State Legislation

Similar efforts are being pursued at the State level with some success. Wyoming and Washington have moved to establish some legal clarity on pore space ownership and have enacted some regulations regarding storage of captured CO_2. Several states have proposed regulations or are in the process of performing assessments and studies in preparation of developing regulations. However, these actions have encountered problems.

California proposed House Bill No. AB 705 for carbon sequestration, but the language was removed from the bill when forwarded to the Chief Clerk on February 1, 2008. Montana proposed Senate Bill No. 562, which was to create a new property class for carbon sequestration pipelines for taxation purposes. The bill did not address CO_2 ownership, ROWs, pore space ownership, or liabilities due to CO_2. Montana Senate Bill No. 218 was introduced to authorize the Board of Environmental Review to adopt rules establishing a CO_2 sequestration program and permit system and providing authority to assess fees, issue penalties, and set bonds. Both of these bills died in the Standing Committee on April 27, 2007.

Washington and Wyoming have put some regulations into place for the operation of long-term CO_2 storage fields. As noted earlier, Wyoming enacted legislation clarifying pore space ownership. Washington's regulations[81] regarding CO_2 injection and storage are similar to those presented by the EPA in its draft rules for Class VI wells. This is an important beginning, but these states need additional legislative work to fully address all the required aspects of CCS projects for both short- and long-term activities.

Addressing Liabilities

Proposed Federal laws are under consideration to resolve the issue of long-term liability; who will be responsible. The "Carbon Capture and Storage Technology Act of 2007," (S.2323) proposes to establish an interagency task force to develop CCS regulations.[82] One of the requirements to be addressed in this pending legislation is the development of regulations regarding the potential and appropriate transfer of liability to government entities. Government entities would be the logical choice since they would exist after demise of the lessee/generator of the carbon dioxide.

Section 8004 of the proposed "Lieberman-Warner Climate Security Act of 2008," (S. 3036) requires the conduct of a study of the legal framework, environmental and safety considerations, and cost implications of potential Federal assumption of liability regarding closed geological storage sites.[83] An Interstate Oil and Gas Compact Commission (IOGCC) task force addressed short- and long-term liabilities in its proposed "Model General Rules and Regulations." The task force reviewed a number of methodologies for insight on proposed ways to develop regulations for liabilities. These included:

- The Texas FutureGen model
- A governmental insurance fund along the lines of the Federal flood insurance program; a private insurance program funded through premiums
- The Price-Anderson Act analog, which would protect the liability of the operator and the CO_2 generators
- The Federal Superfund model under CERCLA
- The Federal Oil Pollution Act of 1990 model acquisition by the state of the storage rights through private purchase of the storage rights from private owners
- The RCRA model where the generators of the CO_2 would be the responsible party

IOGCC concluded that the most efficient methodology to address liability was to utilize the existing frameworks developed by the states for addressing abandoned and orphaned oil and gas wells.[84]

Texas and Illinois have addressed long-term liability on a limited basis for FutureGen but not for other projects. During the initial project selection process, the two candidate states passed into law legislation that would limit operator liabilities associated with the long-term storage of CO_2. Many considered these actions to be a demonstration of commitment to the CCS aspects of the project and a strategic advantage over projects proposed in other states.

During the FutureGen competition, Texas had proposed to accept liability only for FutureGen sites to enhance the chances for Texas to be selected as the location for the project. Illinois also passed legislation to accept liability for the FutureGen sites. The states took an approach to limit their exposure and place stipulations on the circumstances under which they accept the liability. For example, in the case of Illinois, the state would accept liability, but if the contractor was negligent in implementing proper construction practices and codes, the state was not responsible.

Other Storage-Related Regulations and Laws

Additional regulations or laws are worth considering and may affect other aspects of CO_2 storage on Federal land.

BLM has regulations in place that oversee CO_2 production and royalty collection from natural reserves. BLM currently receives a royalty from the production of CO_2 on Federal land.

BLM regulations also address the storage of gas on Federal lands. CO_2 that would be added to a CO_2 production reservoir would need to be accounted for as additional inventory. With current regulations, the CO_2 could later be removed from the reservoir without a royalty. The storage reservoir is subject to rent fees.

Different rules apply to the storage of natural gas and CO_2 on Federal lands. The natural gas is stored on a seasonal basis. The storage of CO_2 under Federal lands will most likely be for the long term. No regulations currently address the long-term storage of CO_2. Currently no information exists on the cost of long-term storage of CO_2 under Federal lands; however, these figures are not expected to be very different from storage under non-Federal lands. Monitoring of the stored CO_2 is also a cost to be paid by the owner or another entity who may own the CO_2.

3.0. CONCLUSION

The total onshore storage resource potential for captured CO_2 in the United States is between 2,289 and 8,129 billion metric tons. The storage resource potential for captured CO_2 beneath Federally managed land within the United States is between 126 to 375 billion metric tons, approximately 5 percent of the estimated national onshore storage potential.

An important initial step in developing a CO_2 storage field is to acquire control of surface acreage that also provides control of the subsurface geologic sink. In assembling an acreage block, Federal lands present two unique advantages, single ownership, and large tracts of land. Trying to secure leases from multiple landowners who may have various opinions and expectations regarding CO_2 storage operations may be quite a challenge. Negotiating with a single landowner to secure the rights to large tracks of land can provide an advantage here, not only for potential future operations but also for early large-scale demonstration projects that will help accelerate commercial deployment of CCS technology.

Clear ownership of the storage reservoir pore space is critical to a successful CO_2 storage project. Surface access for injection facilities and field pipelines as well as for MVA activities is equally important. The area of a CO_2 plume in the subsurface will expand with time and continued injection. Upon cessation of injection, the natural flows of formation waters in saline reservoirs will impart an influence on the plume. These physical attributes must be taken into account when assembling an acreage block many decades before field operations are closed down. Clear ownership of the surface and subsurface establishes the ability to transfer rights for access to a potential storage field operator. This is one advantage Federal lands present to the CCS industry.

To facilitate transportation of captured CO_2, EPACT05 required BLM to perform preliminary work in the area of designating energy corridors on Federal lands for subsurface pipelines. This work includes the preparation of a preliminary EIS. These designated energy corridors will allow placement of CO_2 pipelines and save time and costs for such projects. The timeliness of pipeline projects will be enhanced by the designation of ROW corridors as required by EPACT05 Section 368 and their subsequent inclusion in Federal and state land management plans.

The development of CO_2 pipelines will be adversely impacted by a lack of designated ROW corridors or the absence of ROW needs in State and Federal land-use plans. The ROW permitting process is of concern to getting demonstration and early commercial CCS projects underway because it often requires years to complete and results in high costs.

As stated earlier, Federal lands have two strong advantages, a single owner combined with the availability of large tracts of land, but they come with their own conditions, stipulations, covenants, restrictions, and other obligations. The NEPA requirement is a critical factor in leasing Federal lands. Of the four filing categories under NEPA, an EIS is the most thorough analysis of the proposed situation. Gaining a pipeline ROW through Federal lands requires an EIS and this may become the situation in utilizing a Federal lease for carbon dioxide storage. A National Petroleum Council analysis on the impacts of EIS related surveys on exploration and development drilling activity found that they can add between 1 and 22 months and cost $15,000 to $250,000 for a project.[85]

In addition to EIS, Federal leases may carry seasonal access restrictions due to the effort to protect wildlife and habitat. These restrictions could range from no lease to a time restriction of 3–9 months during which the operator is not allowed access. Furthermore, Federal land leases simply might not be available due to statutory, executive, or administrative actions. Each department or agency might restrict access to all of its managed land or only specific tracts, and therefore the lease for a Federal land parcel would be unavailable.

Federal lands are held for the benefit of the public and leases are available to the public for resource development. For an annual fee, Federal land leases are awarded to the high bidder at a competitive oral auction and Federal land not acquired at the competitive oral auction becomes available afterwards for lease by noncompetitive bidding. Most of these leases convey surface and subsurface rights but in some situations, these rights are severed. Stipulations for use of the land are included in the lease terms and conditions. Use of the Federal land under the lease will involve compliance with applicable Federal, State, and local regulations that are intended to protect the environment and human health. These stipulations may place additional restrictions on access, further reducing net storage potential.

Another barrier that impairs the use of Federal land for CCS activities is the location of major emissions in comparison to the majority of Federal land. Most CO_2 point sources are located east of the Mississippi River, whereas the majority of storage potential beneath Federal land is located west of the Mississippi River.

Because no CCS projects have been permitted on Federal land, no current lease terms and conditions address the ownership of the pore space and storage of CO_2 on the Federal lands. Because Federal land is held in fee simple (except for split estate), the United States owns the surface, subsurface, and any minerals found in either location, including the pore space within which those minerals reside.

Federal leases are designed for mineral extraction. Oil and gas leases provide for production of hydrocarbons that occupy the subsurface pore space. Injecting CO_2 for EOR is one method of production. This will also be the situation for non-Federal leases. Sequestration of CO_2, although similar in many respects to oil and gas operations, is not a production process. Leases that will be used for CCS projects will need to address ownership of the injected CO_2 occupying the subsurface pore space. This is especially important with respect to long-term liability.

The EPA is currently working on CO_2 injection regulations for sequestration. These regulations are currently in draft, and are projected to be in final form in 2011. The rules currently provide an idea of what an operator will be responsible for during injection and post-injection; however certain issues, such as long-term liability, are yet to be addressed and most likely need to be addressed through other regulations.

Currently there are no CO_2 storage laws for Federal land. Existing laws and regulations have analogous requirements for projects, such as those for natural gas, but they require modification and/or additions to accommodate

CO_2 projects. Large-scale demonstration projects are necessary to provide additional information and data for development of CO_2 laws and regulations. States have performed studies to review the effect of modifying analogous laws and regulations to encompass CCS. Studies have also reviewed the effect of new CCS laws and regulations on existing analogous laws and regulations. Currently, states are at different levels when addressing CCS legislation. Some are further along than others, such as Washington and Wyoming. The long-term storage of CO_2 will pose challenges that are new and have uncertainty.

In addition to the laws and regulations, short-term and long-term liabilities must be clearly defined before permanent CO_2 storage beneath Federal lands can be broadly deployed. Short-term liabilities occur during site characterization, construction, injection, closure, and post-closure monitoring and verification phases. The storage field operator (the lessee) would be responsible for operational and environmental liabilities that occur during this period of time. These liabilities are essentially the same as those of oil and gas field operations; however, a major difference will be the level of scrutiny and regulatory oversight.

The purpose of storing captured CO_2 is to permanently prevent it from entering the atmosphere, a goal that represents a very long period of time. Long-term, post-injection liabilities involve leakage and/or migration, which may occur many years or decades (centuries) after cessation of the injection. This issue is tied to the long-term stability of the sequestered CO_2 plume or, as noted earlier, the EPA would define this as a state of non-endangerment. Perhaps even more important here is determining who will be responsible for long-term liability, an intergenerational challenge. A resolution for long-term liability has yet to be agreed upon, although there are several models available from other industries to draw upon. Regardless, either State or Federal involvement is likely. Some options include financial guarantees of performance via surety bonds, collateral bonds, and government- or industry-funded bond pools. Without a transfer of liability upon conclusion of active injection operations, prior to long-term storage, many believe that potential operators will not be willing to enter into a situation from which there is no release.

Long-term storage of CO_2 beneath Federal lands will require long-term monitoring for potential subsurface and surface leakage. Various methods may be utilized for a monitoring system that could be designed for a range of detection levels at a predetermined number of monitoring locations. Currently, no laws or regulations exist to detail these requirements. Without this regulatory framework for long-term monitoring, a CCS project developer would have difficulty in developing proper specifications to ensure long-term injection well integrity and a reliable monitoring system. This also adds to the difficulty of estimating the maintenance and monitoring costs that will need to be provided for in the overall project planning and implementation stages. Without definition of the long-term monitoring requirements, this area of uncertainty raises concern for a prospective CCS project developer. The long-term MVA may have to endure a time range of hundreds to thousands of years.

The absence of regulations governing CO_2 storage operations is the primary obstacle in utilizing the storage potential beneath Federal land. Once regulations are established, the single owner aspect of Federal land will provide an incentive to develop the storage potential beneath these lands. The presence of significant storage potential in Wyoming, Montana and the western Dakotas, in conjunction with a CO_2 pipeline network provides two significant links of the CCS chain. Oil and gas reservoir storage potential in these states represent about 30 percent of the onshore capacity in the Untied States, providing opportunity for CO_2-EOR projects. There are plenty of sources in this area, the third link in the CCS chain, that can utilize this storage potential. Also present is significant coal, oil and natural gas resource potential. Local utilization of these resources, either for electric power generation, CTL/CBTL or natural gas processing can in turn take advantage of the potential storage capacity found in this area. Prudent use of Federal land in this area of the United States can provide coal base load power generation and CTL/CBTL plants, crude oil for refining, natural gas for home and industry and in turn storage of captured CO2 from these operations. But much sooner, once regulations are sorted out, Federal land can provide the necessary acreage and associated storage potential for early deployment of CCS technology.

APPENDIX A: GEOGRAPHIC INFORMATION SYSTEMS

ESRIS's ArcViewGIS 9.2 software was used to generate the maps and data used in this report. The following sections provide a description of the assembly of this data and the procedures used in GIS over the course of this project. All data gathered was current at the time of acquisition in the spring of 2008.

CO_2 Emissions

The initial CO_2 emitting facility spatial data was obtained through NATCARB. This information was used as the CO_2 source data used in the first edition (March, 2007) of NETL's "Carbon Sequestration Atlas of the United States and Canada." Another national layer containing additional ethanol facilities was also obtained from NATCARB. Both layers were accessed in the spring of 2008. Combining these with the most recent regional layers obtained from the individual RCSPs (as listed previously), the data was merged to create a single CO_2 source shapefile. It should be noted that detailed Northeast (no region) data was not readily available because it is not included in a RCSP. Therefore, original national data was used for the Northeast region. After the multiple spatial data sets were obtained, the individual shapefiles were merged to ensure the identification of the greatest number of sources, as well as to ensure the most up-to-date information. The single merged shapefile was then modified to include a unique identifier and coordinates for each point source.

The attributes from the merged shapefile were exported to MS Excel for quality assurance and quality control (QA/QC). The attribute file was tailored by eliminating duplicate sources and non- CO_2 emitting facilities (author discretion was used in that not all information was available for individual sources). Duplicate sites found in the data set using name and identical coordinates were examined individually and compared to release data to determine if there were duplicates or multiple power generation units in the same location. Careful consideration was used in determining duplicates, as it was found that some sites were located based on a city coordinate rather than actual facility coordinate location. In this case, facility name was the determining factor for duplicates. The remaining facilities were then reviewed to update source emission values that originally stated a zero emission. The Center for Global Development – Carbon Monitoring for Action (CARMA) database (2007)[86] and the Ventyx Energy Velocity (EV) database ((2008) license required)[87] were examined for the identified facilities.

Once updated with current emission data and without duplication of any facilities, the spreadsheet was joined with the original merged point source shapefile using the unique identifier previously assigned. This created a new spatial data set for CO_2 point emitters that can be spatially queried against Federal lands.

CO_2 Emission Queries

Several interactive queries were applied to the Region shapefiles, Federal land shapefile and the CO_2 point source spatial data set. An interactive query is a user search in which the user identifies parameters within the data that must be met so the software provides the identified data. The following will describe what parameters were applied to obtain the necessary data.

First, each Federal land category was clipped by all eight U.S. partnership regions (Alaska is included in WESTCARB). The acreages for all Federal land groupings were then calculated within the specified regional area as seen in

Next, a spatial selection query for each partnership was performed to identify the CO_2 emitters producing over 10,000 metric tons of CO_2 per year, both on or within 100 miles of each Federal land group. For example, a single query was run to obtain data that meet all of the following criteria:

a) Federal land that is managed by DOD
b) Land that is within the PCOR region
c) Facilities that produce >10,000 metric tons of CO_2 per year
d) Facilities located on or within 100 miles of DOD managed land

This query would then produce a list of facilities that produce > 10,000 metric tons of CO_2 that are identified as being on or within 100 miles of DOD land located within PCOR. As seen in Table D-1, this list would include 727 facilities.

Lastly, existing natural gas and oil pipeline ROW infrastructures were identified on and within 100 miles of each Federal land group using the same spatial selection query used for CO_2 emitters (region and Federal land group). Sources for ROW infrastructures included Energy Velocity (February 2008) and National Pipeline Mapping System (July 2005). All spatial analysis within GIS was done using projected data assigned with the USA Contiguous Albers Equal Area Conic USGS map projection.

CO_2 Infrastructure

Well Data

A national oil and gas well database was not obtained; however, two regional databases were acquired from the Plains CO_2 Reduction Partnership

(PCOR)[88] and the Midwest Regional Carbon Sequestration Partnership (MRCSP).[89]

Apart from the regional data, a state level search was the major focal point for well information and location data. The key focus of well data was current oil and gas producing states such as Alabama, Alaska, Arizona, Arkansas, California, Colorado, Florida, Illinois, Indiana, Kansas, Kentucky, Louisiana, Michigan, Mississippi, Missouri, Montana, Nebraska, Nevada, New Mexico, New York, North Dakota, Ohio, Oklahoma, Pennsylvania, South Dakota, Tennessee, Texas, Utah, Virginia, West Virginia, and Wyoming. Well data was initially collected through an internet search of state Departments and Agencies. If internet searches were ineffective, direct contact was used to obtain data. Well data and information sources can be seen in Table A-1.

Not all states from the previously listed oil and gas producing states provided readily available information. Some state data was not easily accessed or obtained through internet searches or contacts (Arizona, Missouri, Tennessee, and Texas). This could simply be because electronic data is not made available to the public without a cost.

Once state well data was collected it was merged into a master well file. Because well data came in different file types such as ESRI Geodatabases, ESRI shapefiles, MS Excel spreadsheets (xls), MS Access databases (mdb), and databases (dbf), it was critical to merge the information into one easily accessible format. First, raw data was converted or exported to ESRI shapefile format using provided coordinates such as latitude-longitude or Universal Transverse Mercator (UTM). Each state shapefile was then queried for wells on and within 100 miles of Federal lands to eliminate wells outside the 100 mile Federal land buffer. Queried subsets were re-projected to *USA Contiguous Albers Equal Area Conic USGS version* projection for use in regional analysis. New coordinates reflecting this projection were calculated for each well location.

Re-projected subset state well files were exported to dbase format and incorporated into MS Access database tables to be cleaned of unwanted data and column formatting. Each state file included a vast range of wells and supporting information for each well. It was determined that only the following identified characteristics for each well would be kept: American Petroleum Institute (API) number, well name, operator name, well type, well status, completion date, plug date, spud date, permit date, well elevation, total measured depth, total vertical depth, field name, surface latitude, surface longitude, and state location of well.

Table A-1. Oil and Gas Well Data Sources

STATE	SOURCE	WEBSITE
Alabama	Alabama Oil and Gas Board	http://www.ogb.state.al.us/
Alaska	Alaska Division of Oil and Natural Gas	http://www.dog.dnr.state.ak.us/oil/products/data/wells/wells.htm
Arkansas	Arkansas GeoStor 5.0	http://www.geostor.arkansas.gov/Portal/index.jsp
California	California Department of Conservation	http://www.consrv.ca.gov/dog/maps/Pages/goto_welllocation.aspx
Colorado	Colorado Oil and Gas Conservation Commission	http://oil-gas.state.co.us/infosys/Maps/gismain.cfm#Downloads
Florida	Florida Department of Environmental Protection	http://www.dep.state.fl.us/geology/gisdatamaps/oil_gas_permit_data.h tm #shapefile
Illinois	Illinois Natural Resources Geospatial Data Clearinghouse	http://www.isgs.uiuc.edu/nsdihome/webdocs/st-geolb.html#wells
Indiana	Indiana Geological Survey GIS Atlas	http://129.79.145.7/arcims/statewide_mxd/download.html
Kansas	Kansas Geological Survey	http://www.kgs.ku.edu/PRS/petroDB.html
Kentucky	University of Kentucky - Kentucky Geological Survey	http://www.uky.edu/KGS/emsweb/data/kyogshape.html
Louisiana	Louisiana Statewide GIS Atlas	http://atlas.lsu.edu/search/default.asp
Michigan	Michigan Department of Environmental Quality	http://www.michigan.gov/deq/0,1607,7-135-3311_4111_4231-97870--,00.html
Mississippi	Mississippi Automated Resource Information System	http://www.maris.state.ms.us/HTM/DownloadData/Statewide-Alpha.htm

Table A-1. (Continued)

STATE	SOURCE	WEBSITE
Montana	DNRC: Montana Board of Oil and Gas	http://bogc.dnrc.state.mt.us/jdpintro.asp
Nebraska	Nebraska Oil and Gas Conservation Commission	http://www.nogcc.ne.gov/NOGCCPublications.htm
Nevada	Nevada Bureau of Mines and Geology	http://www.nbmg.unr.edu/lists/oil/oil.htm
New Mexico	Petroleum Recovery Research Center	http://ocdimage.emnrd.state.nm.us/imaging/WellFileCriteria.aspx
New York	New York Department of Environmental Conservation	http://www.dec.ny.gov/energy/1603.html
North Dakota	North Dakota Industrial Commission Department of Mineral Resources	https://www.dmr.nd.gov/oilgas/
Ohio	Ohio Geological Survey	http://www.ohiodnr.com/geosurvey/ogcim/petrol/digmaps/tabid/7773/Default.aspx
Oklahoma	Oklahoma Corporation Commission	http://www.occ.state.ok.us/Divisions/OG/newweb/ogdatafiles.htm
Pennsylvania	Pennsylvania Spatial Data Access	http://www.pasda.psu.edu/
South Dakota	South Dakota Department of Environmental and Natural Resources	http://www.state.sd.us/denr/DES/Mining/Oil&Gas/producti.htm
Utah	Utah Division of Oil, Gas, and Mining	http://oilgas.ogm.utah.gov/Data_Center/DataCenter.cfm#download

STATE	SOURCE	WEBSITE
Virginia	Virginia Department of Mines, Minerals, and Energy	http://www.dmme.virginia.gov/trans/DgoFiles.xls
West Virginia	West Virginia Department of Environmental Protection	http://www.wvdep.org/item.cfm?ssid=23&ss1id=97
Wyoming	Wyoming Oil and Gas Conservation Commission	http://wogcc.state.wy.us/
MRCSP	Ohio Geological Survey - MRCSP	ftp://ftp.dnr.state.oh.us/Geological_Survey/MRCSP
PCOR	PCOR ftp site provided by PCOR	ftp://ftp.undeerc.org/PCOR/

The select attributes were then identified, labeled, and organized. Excess information was deleted. Each table was formatted with the same column fields and field type in preparation to be merged into one file. Minimal data loss occurred during the data field type changes.

This process created a master wells database which included a table for each identified state and region. The tables were then exported as database files (dbfs) and individual shapefiles were created using projected coordinates. The resulting shapefiles were then merged into one master well shapefile.

CO_2 Pipeline Data

An ESRI shapefile of U.S. CO_2 pipelines provided by National Carbon Sequestration Database and Geographic Information System (NATCARB)[90] was used to identify existing and proposed CO_2 pipelines across the United States.

Natural Gas and Petroleum Pipeline Data

ESRI shapefiles of natural gas and petroleum pipelines were obtained through Ventyx Energy Velocity (EV) database (February 2008)91 and National Pipeline Mapping System (NPMS) database (July 2005).[92]

Queries

Well Queries

First, each Federal land category was clipped by all eight U.S. partnership regions (Alaska is included in WESTCARB). Each Federal land group within the regions was then selected to be used as criteria. Next, two spatial selection queries for each partnership and Federal land group were performed to identify the oil and gas wells on or within 100 miles of each Federal land group. For example, a single query was run to obtain data that meet all of the following criteria:

a) Federal land that is managed by FWS
b) Land that is within the MRCSP region
c) Oil and gas wells located on or within 100 miles of managed land

This query would then produce a list of wells that are identified as being on or within 100 miles of FWS land located within MRCSP. As seen in Table D-8, this list includes 390,790 wells.

Next, a spatial selection query for each partnership was performed to identify the oil and gas wells on each Federal land group. For example, a single query was run to obtain data that meet all of the following criteria:

a) Federal land that is managed by DOE
b) Land that is within the MGSC region
c) Oil and gas wells located on DOE managed land

This query would then produce a list of wells that are identified as being on DOE land located within MGSC. As seen in Table D-8, this list includes 356 wells.

All query information was compiled to generate the total number of wells for each land group and region (Table D-7). Note that a single well might be within range of multiple Federal land groups and, therefore, included more than once (data summary driven by regions).

For example, within the SWP region, a well might be located on or within 100 miles of NPS land; however, that same well might be on or within 100 miles of FS land.

Therefore, that well would be included when summing the total number of oil and gas wells on or within 100 miles of both of those Federal agencies. Consequently, the "Total" column located in Table D-7 is not a sum of the Group columns; it is simply the total within a region. Table D-8 "Total" column, however, is an actual total for each region because a well cannot be located on multiple Federal lands.

CO_2 Pipeline Queries

CO_2 pipelines on or within 100 miles of Federal lands were identified using a 100 mile buffer selection query against the USGS Federal lands shapefile. All existing CO_2 pipelines fell into this category.

Geologic Storage Potential

Regional Partnership Data

Each of the seven RCSPs was tasked with providing CO_2 capacity estimates for the identified geologic formations: saline formations, oil and gas reservoirs, and unmineable coal seams. Each of the RCSPs information was collected and analyzed for the quantification of storage resources available at a subregional scale.

Each partnership was directed to follow a common methodology for capacity estimates and data reporting (methodology can be found in the DOE *Sequestration Atlas of the United States and Canada*). This data was then submitted to NATCARB for compiling and manipulation to a single format.

NATCARB Data

Once received, NATCARB converted each RCSP data set into a common format. Each geologic formation file was merged into a single sink layer representing the entire United States. For example, all individual oil and gas reservoir files were merged into one layer encompassing all of the oil and gas reservoir locations and potential storage capacity within the United States. In addition to merging, any sink capacity overlap between RCSPs were identified and averaged.

The distribution of sinks representing potential CO_2 storage capacity throughout the continental United States and Alaska were provided by NATCARB in shapefile format for the three identified geologic sinks: saline formations, oil and gas reservoirs, and unmineable coal seams. The resource value, the number that represents the estimated storage volume for any one of these sinks, is a summary value for each of the three potential storage horizons provided by each RCSP.

The storage potential of a reservoir in a specific geologic horizon (i.e. the Mt. Simon sandstone, Madison Limestone or Pittsburgh Coal) within the overall stratigraphic column in a particular region is not available from this database.

Clips, Queries, and CO_2 Storage Capacity Calculations

Potential CO_2 storage sinks throughout the continental U.S. and Alaska were provided by NATCARB in ESRI shapefile format for the three identified geologic sinks: saline formations, oil and gas reservoirs, and unmineable coal seams.

The following was applied to each national geologic layer. First, a definition query was used to filter out sinks not assessed and/or deemed unsuitable for CO_2 storage by RCSP. Sinks without storage potential vary from region to region depending on each RCSP criteria for characterizing sinks.

Next, new fields were added to each layer using the "add field" command within the "options" menu of the layer attribute table.

The new fields added included total sink surface area in acres ("ACRES"), sink surface area in acres located below Federal land ("FED_ACRES"), percentage of sinks located below Federal land ("FED_PERC"), low storage potential in metric tons beneath Federal land ("FED_L_CAP"), and high storage potential in metric tons beneath Federal land ("FED_H_CAP"). Once added, the "ACRES" field was populated using the "Calculate geometry" command for acres using ESRI's "Field Calculator" within the layers attribute table. This established an original surface area to base storage potential on when examining Federal land subset areas.

The remaining fields were then calculated after each of the three the national layers was "clipped" by RCSP Federal land group, resulting in subset shapefiles for each Federal land group in each RCSP. The "FED_ACRES" field was then calculated using the "Calculate geometry" command for acres using the "Field calculator" to determine the subset acreages.

Next, the percentage of Federal land sink area was calculated by dividing "FED_ACRES" by the original "ACRES" field and the result was used to populate the "FED_PERC" field. These percentages were then multiplied by the original total low and high sink storage potential estimates for the appropriate sink and resulting values were then used to populate the "FED_L_CAP" and "FED_H_CAP" fields.

This process was performed for all three national layers within each of the eight partnership regions to calculate gross storage potential (see Tables D-5 and D-6, Appendix D).

Finally, this process was repeated for all three national layers within each of the eight partnership regions to obtain storage potential beneath leasable Federal land.

Appendix B: Federal Land Groups from Initial USGS Groups

GROUP 1: DOD	GROUP 2: DOE	GROUP 3: BLM	GROUP 4: BOR	GROUP 5: FWS
Air Force DOD	Department of Energy DOE	Forest Reserve BLM	Bureau of Reclamation BOR	National Fish Hatchery FWS
Army Corps of Engineers DOD		National Conservation Area BLM		National Preserve FWS
Army DOD		National Monument BLM		National Wildlife Refuge FWS
Department of Defense DOD		National Recreation Area BLM		Waterfowl Production Area FWS
Marine Corps DOD		National Wild and Scenic River BLM		Wilderness FWS
Navy DOD		Other BLM		Wilderness Study Area FWS
		Public Domain Land BLM		Wildlife Management Area FWS
		Wilderness BLM		
		Wilderness Study Area BLM		

GROUP 6: NPS	GROUP 7: DOJ	GROUP 8: Other Federal Agencies	GROUP 9: FS and Other USDA
National Battlefield NPS	Bureau of Prisons DOJ	Central Intelligence Agency CIA	National Forest FS
National Battlefield Park NPS		General Services Administration GSA	National Grassland FS
National Capital Park NPS		National Aeronautics Administration NASA	National Monument FS
National Historic Landmark District NPS		Metropolitan Washington Airports Authority	National Recreation Area FS
National Historic Park NPS		TVA	National Scenic Area FS
National Historic Site NPS		U.S. Coast Guard DOT	Purchase Unit Block FS
National Lakeshore NPS		U.S. Department of Transportation DOT	Wilderness FS
National Mall NPS			Wilderness Study Area FS
National Memorial NPS			Agricultural Research Service ARS
National Military Park NPS			United States Department of Agriculture USDA
National Monument NPS			
National Park NPS			
National Parkway NPS			
National Preserve NPS			
National Recreation Area NPS			
National Reserve NPS			
National River NPS			
National Scenic River NPS			
National Seashore NPS			
National Wild and Scenic River NPS			
Wilderness NPS			
Wilderness Study Area NPS			

Appendix C. Federal Land Access Categorization

No leasing (some exceptions may exist)

GROUP 1: DOD	GROUP 3: BLM	GROUP 4: BOR
Air Force DOD	National Conservation Area BLM	National Fish Hatchery FWS
Army Corps of Engineers DOD	National Monument BLM	National Preserve FWS
Army DOD	National Recreation Area BLM	National Wildlife Refuge FWS
Department of Defense DOD	National Wild and Scenic River BLM	Waterfowl Production Area FWS
Marine Corps DOD	Wilderness BLM	Wilderness FWS
Navy DOD	Wilderness Study Area BLM	Wilderness Study Area FWS
GROUP 6: NPS	GROUP 9: FS and Other USDA	Wildlife Management Area FWS
National Battlefield NPS	National Monument FS	
National Battlefield Park NPS	National Recreation Area FS	
National Capital Park NPS	National Scenic Area FS	
National Historic Landmark District NPS	Wilderness FS	
National Historic Park NPS	Wilderness Study Area FS	
National Historic Site NPS		
National Lakeshore NPS		
National Mall NPS		
National Memorial NPS		
National Military Park NPS		
National Monument NPS		
National Park NPS		
National Parkway NPS		
National Preserve NPS		
National Recreation Area NPS		

(Continued)

GROUP 1: DOD	GROUP 3: BLM	GROUP 4: BOR
National Reserve NPS		
National River NPS		
National Scenic River NPS		
National Seashore NPS		
National Wild and Scenic River NPS		
Wilderness NPS		
Wilderness Study Area NPS		

Leasing (subject to stipulations)

GROUP 2: DOE	GROUP 3: BLM	GROUP 4: BOR
Department of Energy DOE	Forest Reserve BLM	Bureau of Reclamation BOR
	Other BLM	
	Public Domain Land BLM	
GROUP 7: DOJ	GROUP 8: Other Federal Agencies	GROUP 9: FS and Other USDA
Bureau of Prisons DOJ	U.S. Coast Guard DOT	National Forest FS
	U.S. Department of Transportation DOT	National Grassland FS
		Purchase Unit Block FS
		United States Department of Agriculture USDA

Leasing status unknown

GROUP 8: Other Federal Agencies	GROUP 9: FS and Other USDA
Central Intelligence Agency CIA	Agricultural Research Service ARS
General Services Administration GSA	
National Aeronautics Administration NASA	
Metropolitan Washington Airports Authority	
TVA	

APPENDIX D: TOTAL FEDERAL LAND DATA (LEASABLE AND NON-LEASABLE)

Table D-1. CO_2 Emitting Facilities (> 10,000 Metric Tons/Year) On and Within 100 Miles of Federal Land Group

		Group 1 Totals	Group 2 Totals	Group 3 Totals	Group 4 Totals	Group 5 Totals	Group 6 Totals	Group 7 Totals	Group 8 Totals	Group 9 Totals
RCSP REGION	REGION TOTALS	DOD	DOE	BLM	BOR	FWS	NPS	DOJ	Other Federal Agencies	FS and Other USDA
PCOR	1,037	727	0	143	157	893	439	55	0	606
MGSC	560	528	222	0	0	413	66	90	62	348
MRCSP	870	668	128	0	0	488	596	0	67	494
WESTCARB	137	6	13	113	81	104	114	0	16	130
SECARB	898	865	52	0	140	763	713	85	263	707
SWP	467	360	3	202	245	342	219	29	34	290
Big Sky	236	140	13	230	197	161	236	0	0	219
Northeast	86	268	0	0	0	268	231	117	0	86

Table D-2. CO_2 Emissions (Million Metric Tons/Year) Produced On and Within 100 Miles of Federal Land Group

		Group 1 Totals	Group 2 Totals	Group 3 Totals	Group 4 Totals	Group 5 Totals	Group 6 Totals	Group 7 Totals	Group 8 Totlas	Group 9 Totals
RCSP REGION	REGION TOTALS	DOD	DOE	BLM	BOR	FWS	NPS	DOJ	Other Federal Agencies	FS and Other USDA
PCOR	732	509	0	131	126	638	330	64	0	434
MGSC	856	840	285	0	0	675	187	165	228	677
MRCSP	1,156	1,058	61	0	0	725	863	0	88	795
WESTCARB	203	4	17	170	101	117	191	0	22	197
SECARB	1,293	1,244	91	0	150	1,051	966	144	500	1,020
SWP	570	450	1	315	309	343	342	30	31	406
Big Sky	158	74	3	156	106	96	158	0	0	148
Northeast	40	188	0	0	0	189	147	58	0	40

Table D-3. Oil and Gas Wells Located On and Within 100 Miles of Federal Land Group

		Group 1 Totals	Group 2Totals	Group 3 Totals	Group 4 Totals	Group 5 Totals	Group 6 Totals	Group 7 Totals	Group 8 Totals	Group 9 Totals
RCSP REGION	REGION TOTALS	DOD	DOE	BLM	BOR	FWS	NPS	DOJ	Other Federal Agencies	FS and Other USDA
PCOR	877,773	419,042	0	192,031	249,014	641,685	227,917	58,270	0	411,275
MGSC	750,309	748,741	327,832	0	0	605,943	103,743	225,362	141,478	596,311
MRCSP	818,438	760,351	486	0	0	390,790	700,700	0	230	578,399
WESTCARB	197,374	195,778	546	197,311	189,082	195,666	193,692	0	0	191,557
SECARB	671,401	585,787	38,684	0	0	331,332	393,963	100	231,257	553,786
SWP	921,703	797,502	512	160,623	654,453	838,309	521,380	38,427	40,983	454,159
Big Sky	256,587	182,883	71	252,372	202,771	94,998	82,635	0	0	233,079
Northeast	519	315	0	0	0	374	508	239	0	73

Table D-4. Oil and Gas Wells Located on Federal Land Group

RCSP REGION	TOTAL Of Groups By RCSP	Group 1 Totals	Group 2 Totals	Group 3 Totals	Group 4 Totals	Group 5 Totals	Group 6 Totals	Group 7 Totals	Group 8 Totals	Group 9 Totals
		DOD	DOE	BLM	BOR	FWS	NPS	DOJ	Other Federal Agencies	FS and Other USDA
PCOR	27,034	167	0	9,120	36	155	3	0	0	17,553
MGSC	5,802	1,177	356	0	0	229	10	13	3	4,014
MRCSP	53,257	1,322	0	0	0	33	344	0	0	51,558
WESTCARB	17,053	6,728	0	8,371	21	969	69	0	0	895
SECARB	8,277	886	0	0	0	3,060	170	0	0	4,161
SWP	30,386	3,493	0	20,894	134	566	121	10	1	5,167
Big Sky	37,223	1,654	0	34,015	1,160	2	4	0	0	388
Northeast	0	0	0	0	0	0	0	0	0	0

Table D-5. Storage Potential Beneath Federal Land (High Estimate-Million Metric Tons)

RCSP REGION	TOTAL of Groups by RCSP	Group 1 Totals	Group 2 Totals	Group 3 Totals	Group 4 Totals	Group 5 Totals	Group 6 Totals	Group 7 Totals	Group 8 Totals	Group 9 Totals
		DOD	DOE	BLM	BOR	FWS	NPS	DOJ	Other Federal Agencies	FS and Other USDA
PCOR	177,672	1,631	0	104,824	527	11,335	559	0	0	58,795
MGSC	5,530	832	0	0	0	239	0	3	0	4,456
MRCSP	10,117	533	2	0	0	80	252	0	13	9,237
WESTCARB	107,958	17,611	0	30,458	144	8,269	1,871	0	1	49,606
SECARB	249,909	27,377	12	0	1,217	68,494	21,062	4	1,402	130,340
SWP	35,362	18	0	20,010	195	12	19	0	3	15,105
Big Sky	351,673	934	0	284,176	20,322	651	4,247	0	0	41,343
Northeast	0	0	0	0	0	0	0	0	0	0
Total*	938,222	48,936	15	439,467	22,405	89,080	28,010	7	1,418	308,883

* sum may not equal due to rounding error

Table D-6. Storage Potential Beneath Federal Land (Low Estimate-Million Metric Tons)

RCSP REGION	TOTAL of Groups by RCSP	Group 1 Totals	Group 2 Totals	Group 3 Totals	Group 4 Totals	Group 5 Totals	Group 6 Totals	Group 7 Totals	Group 8 Totals	Group 9 Totals
		DOD	DOE	BLM	BOR	FWS	NPS	DOJ	Other Federal Agencies	FS and Other USDA
PCOR	66,496	734	0	32,898	134	2,852	178	0	0	29,701
MGSC	1,429	238	0	0	0	76	0	1	0	1,114
MRCSP	10,100	533	2	0	0	80	246	0	4	9,236
WESTCARB	27,118	4,411	0	7,675	36	2,071	467	0	0	12,457
SECARB	67,259	7,707	3	0	315	17,801	5,346	1	354	35,733
SWP	9,096	6	0	5,140	52	5	5	0	1	3,888
Big Sky	92,945	824	0	75,235	5,111	163	1,062	0	0	10,550
Northeast	0	0	0	0	0	0	0	0	0	0
Total*	274,444	14,453	6	120,947	5,648	23,048	7,303	2	358	102,679

* sum may not equal due to rounding error

Table D-7. Oil and Gas Wells Located On and Within 100 Miles of Leasable Federal Land

		Group 1 Total	Group 2 Total	Group 3 Total	Group 4 Total	Group 5 Total	Group 6 Total	Group 7 Total	Group 8 Total	Group 9 Total
RCSP REGION	REGION TOTAL	DOD	DOE	BLM	BOR	FWS	NPS	DOJ	Other Federal Agencies	FS and other USDA
PCOR	509,996	0	0	192,031	249,014	0	0	58,270	0	400,873
MGSC	601,105	0	327,832	0	0	0	0	225,362	0	596,311
MRCSP	578,653	0	486	0	0	0	0	0	0	578,396
WESTCARB	197,315	0	546	197,311	189,082	0	0	0	0	191,557
SECARB	553,828	0	38,684	0	0	0	0	100	11	553,786
SWP	769,210	0	512	160,623	654,453	0	0	38,427	12,324	454,153
Big Sky	252,853	0	71	252,372	202,771	0	0	0	0	674
Northeast	271	0	0	0	0	0	0	239	0	73

Table D-8. Oil and Gas Wells Located on Leasable Federal Land

RCSP REGION	TOTAL of Groups by RCSP	Group 1 Totals	Group 2 Totals	Group 3 Totals	Group 4 Totals	Group 5 Totals	Group 6 Totals	Group 7 Totals	Group 8 Totals	Group 9 Totals
		DOD	DOE	BLM	BOR	FWS	NPS	DOJ	Other Federal Agencies	FS and other USDA
PCOR	26,656	0	0	9,070	36	0	0	0	0	17,550
MGSC	4,352	0	356	0	0	0	0	13	0	3,983
MRCSP	51,397	0	0	0	0	0	0	0	0	51,397
WESTCARB	5,064	0	0	4,221	21	0	0	0	0	822
SECARB	4,154	0	0	0	0	0	0	0	0	4,154
SWP	25,492	0	0	20,186	134	0	0	10	1	5,161
Big Sky	35,175	0	0	34,015	1,160	0	0	0	0	0
Northeast	0	0	0	0	0	0	0	0	0	0

APPENDIX E: FEDERAL LAND SALINE FORMATION DATA

Table E-1. Storage Potential for Deep Saline Formations Beneath Federal Land (High Estimate – Million Metric Tons)

RCSP REGION	TOTAL of Groups by RCSP	Group 1 Totals	Group 2 Totals	Group 3 Totals	Group 4 Totals	Group 5 Totals	Group 6 Totals	Group 7 Totals	Group 8 Totals	Group 9 Totals
		DOD	DOE	BLM	BOR	FWS	NPS	DOJ	Other Federal Agencies	FS and Other USDA
PCOR	148,303	1,201	0	95,930	525	11,316	509	0	0	38,821
MGSC	5,512	816	0	0	0	238	0	3	0	4,455
MRCSP	7,185	501	2	0	0	76	200	0	13	6,392
WESTCARB	107,958	17,611	0	30,458	144	8,269	1,871	0	1	49,606
SECARB	240,109	25,757	12	0	1,196	67,209	20,917	4	1,394	123,620
SWP	35,037	13	0	19,829	194	9	16	0	3	14,973
Big Sky	347,206	147	0	280,556	20,274	651	4,247	0	0	41,332
Northeast	0	0	0	0	0	0	0	0	0	0
Total*	891,309	46,045	15	426,774	22,332	87,768	27,759	7	1,410	279,199

* sum may not equal due to rounding error

Table E-2. Table Storage Potential for Deep Saline Formations Beneath Federal Land
(Low Estimate – Million Metric Tons)

RCSP REGION	TOTAL of Groups by RCSP	Group 1 Totals	Group 2 Totals	Group 3 Totals	Group 4 Totals	Group 5 Totals	Group 6 Totals	Group 7 Totals	Group 8 Totals	Group 9 Totals
		DOD	DOE	BLM	BOR	FWS	NPS	DOJ	Other Federal Agencies	FS and Other USDA
PCOR	37,127	304	0	24,005	132	2,833	127	0	0	9,726
MGSC	1,417	226	0	0	0	76	0	1	0	1,114
MRCSP	7,169	501	2	0	0	76	193	0	4	6,392
WESTCARB	27,118	4,411	0	7,675	36	2,071	467	0	0	12,457
SECARB	60,030	6,439	3	0	299	16,802	5,230	1	349	30,907
SWP	8,988	3	0	5,094	52	2	4	0	1	3,832
Big Sky	88,666	37	0	71,784	5,076	163	1,062	0	0	10,545
Northeast	0	0	0	0	0	0	0	0	0	0
Total*	230,515	11,922	6	108,558	5,594	22,024	7,084	2	353	74,973

* sum may not equal due to rounding error

Table E-3. Storage Potential for Deep Saline Formations Beneath Leasable Federal Land (High Estimate – Million Metric Tons)

RCSP REGION	TOTAL of Groups by RCSP	Group 1 Totals	Group 2 Totals	Group 3 Totals	Group 4 Totals	Group 5 Totals	Group 6 Totals	Group 7 Totals	Group 8 Totals	Group 9 Totals
		DOD	DOE	BLM	BOR	FWS	NPS	DOJ	Other Federal Agencies	FS and Other USDA
PCOR	32,844	0	0	6,902	525	0	0	0	0	25,417
MGSC	4,458	0	0	0	0	0	0	3	0	4,455
MRCSP	6,396	0	2	0	0	0	0	0	0	6,394
WESTCARB	35,910	0	0	16,017	144	0	0	0	1	19,749
SECARB	34,529	0	12	0	1,196	0	0	4	0	33,318
SWP	34,577	0	0	19,824	194	0	0	0	3	14,557
Big Sky	188,048	0	0	167,420	20,274	0	0	0	0	355
Northeast	0	0	0	0	0	0	0	0	0	0
Total*	336,763	0	15	210,162	22,332	0	0	7	4	104,244

* sum may not equal due to rounding error

Table E-4. Storage Potential for Deep Saline Formations Beneath Leasable Federal Land (Low Estimate – Million Metric Tons)

		Group 1 Totals	Group 2 Totals	Group 3 Totals	Group 4 Totals	Group 5 Totals	Group 6 Totals	Group 7 Totals	Group 8 Totals	Group 9 Totals
RCSP REGION	TOTAL of Groups by RCSP	DOD	DOE	BLM	BOR	FWS	NPS	DOJ	Other Federal Agencies	FS and Other USDA
PCOR	8,245	0	0	1,738	132	0	0	0	0	6,375
MGSC	1,115	0	0	0	0	0	0	1	0	1,114
MRCSP	6,395	0	2	0	0	0	0	0	0	6,392
WESTCARB	9,066	0	0	4,056	36	0	0	0	0	4,973
SECARB	8,633	0	3	0	299	0	0	1	0	8,330
SWP	8,868	0	0	5,093	52	0	0	0	1	3,723
Big Sky	48,152	0	0	42,980	5,076	0	0	0	0	96
Northeast	0	0	0	0	0	0	0	0	0	0
Total*	90,472	0	6	53,867	5,594	0	0	2	1	31,003

* sum may not equal due to rounding error

APPENDIX F: FEDERAL LAND OIL AND GAS RESERVOIR DATA

Table F-1. Storage Potential for Oil and Gas Reservoirs Beneath Federal Land (High Estimate – Million Metric Tons)

RCSP REGION	TOTAL of Groups by RCSP	Group 1 Totals	Group 2 Totals	Group 3 Totals	Group 4 Totals	Group 5 Totals	Group 6 Totals	Group 7 Totals	Group 8 Totals	Group 9 Totals
		DOD	DOE	BLM	BOR	FWS	NPS	DOJ	Other Federal Agencies	FS and Other USDA
PCOR	26,095	430	0	6,650	2	19	36	0	0	18,956
MGSC	1	0	0	0	0	0	0	0	0	0
MRCSP	2,858	28	0	0	0	3	50	0	0	2,776
WESTCARB	0	0	0	0	0	0	0	0	0	0
SECARB	1,301	437	0	0	2	328	46	0	0	489
SWP	1	0	0	1	0	0	0	0	0	0
Big Sky	4,089	787	0	3,269	30	0	0	0	0	3
Northeast	0	0	0	0	0	0	0	0	0	0
Total*	34,344	1,683	0	9,920	34	351	133	0	0	22,224

* sum may not equal due to rounding error

Table F-2. Storage Potential for Oil and Gas Reservoirs Beneath Federal Land (Low Estimate – Million Metric Tons)

RCSP REGION	TOTAL of Groups by RCSP	Group 1 Totals	Group 2 Totals	Group 3 Totals	Group 4 Totals	Group 5 Totals	Group 6 Totals	Group 7 Totals	Group 8 Totals	Group 9 Totals
		DOD	DOE	BLM	BOR	FWS	NPS	DOJ	Other Federal Agencies	FS and Other USDA
PCOR	26,095	430	0	6,650	2	19	36	0	0	18,956
MGSC	0	0	0	0	0	0	0	0	0	0
MRCSP	2,858	28	0	0	0	3	50	0	0	2,776
WESTCARB	0	0	0	0	0	0	0	0	0	0
SECARB	1,301	437	0	0	2	328	46	0	0	489
SWP	1	0	0	1	0	0	0	0	0	0
Big Sky	4,089	787	0	3,269	30	0	0	0	0	3
Northeast	0	0	0	0	0	0	0	0	0	0
Total*	34,344	1,683	0	9,920	34	350	133	0	0	22,224

* sum may not equal due to rounding error

Table F-3. Storage Potential for Oil and Gas Reservoirs Beneath Leasable Federal Land (High Estimate – Million Metric Tons)

	TOTAL of Groups by RCSP	Group 1 Totals	Group 2 Totals	Group 3 Totals	Group 4 Totals	Group 5 Totals	Group 6 Totals	Group 7 Totals	Group 8 Totals	Group 9 Totals
RCSP REGION		DOD	DOE	BLM	BOR	FWS	NPS	DOJ	Other Federal Agencies	FS and Other USDA
PCOR	25,416	0	0	6,458	2	0	0	0	0	18,956
MGSC	0	0	0	0	0	0	0	0	0	0
MRCSP	2,769	0	0	0	0	0	0	0	0	2,769
WESTCAR B	0	0	0	0	0	0	0	0	0	0
SECARB	490	0	0	0	2	0	0	0	0	488
SWP	1	0	0	1	0	0	0	0	0	0
Big Sky	3,153	0	0	3,123	30	0	0	0	0	0
Northeast	0	0	0	0	0	0	0	0	0	0
Total*	31,829	0	0	9,582	34	0	0	0	0	22,213

* sum may not equal due to rounding error

Table F-4. Storage Potential for Oil and Gas Reservoirs Beneath Leasable Federal Land (Low Estimate – Million Metric Tons)

RCSP REGION	TOTAL of Groups by RCSP	Group 1 Totals	Group 2 Totals	Group 3 Totals	Group 4 Totals	Group 5 Totals	Group 6 Totals	Group 7 Totals	Group 8 Totals	Group 9 Totals
		DOD	DOE	BLM	BOR	FWS	NPS	DOJ	Other Federal Agencies	FS and Other USDA
PCOR	25,416	0	0	6,458	2	0	0	0	0	18,956
MGSC	0	0	0	0	0	0	0	0	0	0
MRCSP	2,769	0	0	0	0	0	0	0	0	2,769
WESTCAR B	0	0	0	0	0	0	0	0	0	0
SECARB	490	0	0	0	2	0	0	0	0	488
SWP	1	0	0	1	0	0	0	0	0	0
Big Sky	3,153	0	0	3,123	30	0	0	0	0	0
Northeast	0	0	0	0	0	0	0	0	0	0
Total*	31,829	0	0	9,582	34	0	0	0	0	22,213

* sum may not equal due to rounding error

APPENDIX G: FEDERAL LAND COAL SEAM DATA

Table G-1. Storage Potential for Unmineable Coal Seams Beneath Federal Land (High Estimate – Million Metric Tons)

RCSP REGION	TOTAL of Groups by RCSP	Group 1 Totals	Group 2 Totals	Group 3 Totals	Group 4 Totals	Group 5 Totals	Group 6 Totals	Group 7 Totals	Group 8 Totals	Group 9 Totals
		DOD	DOE	BLM	BOR	FWS	NPS	DOJ	Other Federal Agencies	FS and Other USDA
PCOR	3,275	0	0	2,243	0	0	14	0	0	1,018
MGSC	17	16	0	0	0	0	0	0	0	1
MRCSP	75	3	0	0	0	0	2	0	0	69
WESTCARB	0	0	0	0	0	0	0	0	0	0
SECARB	8,499	1,183	0	0	19	957	100	0	8	6,232
SWP	325	5	0	180	1	3	3	0	0	132
Big Sky	378	0	0	351	19	1	0	0	0	8
Northeast	0	0	0	0	0	0	0	0	0	0
Total*	12,569	1,208	0	2,774	39	961	119	0	8	7,460

* sum may not equal due to rounding error

Table G-2. Storage Potential for Unmineable Coal Seams Beneath Federal Land
(Low Estimate – Million Metric Tons)

RCSP REGION	TOTAL of Groups by RCSP	Group 1 Totals	Group 2 Totals	Group 3 Totals	Group 4 Totals	Group 5 Totals	Group 6 Totals	Group 7 Totals	Group 8 Totals	Group 9 Totals
		DOD	DOE	BLM	BOR	FWS	NPS	DOJ	Other Federal Agencies	FS and Other USDA
PCOR	3,275	0	0	2,243	0	0	14	0	0	1,018
MGSC	12	11	0	0	0	0	0	0	0	1
MRCSP	73	3	0	0	0	0	2	0	0	68
WESTCARB	0	0	0	0	0	0	0	0	0	0
SECARB	5,928	831	0	0	13	671	70	0	6	4,337
SWP	107	3	0	45	0	2	1	0	0	56
Big Sky	190	0	0	181	6	0	0	0	0	2
Northeast	0	0	0	0	0	0	0	0	0	0
Total*	9,585	848	0	2,469	20	674	87	0	6	5,482

* sum may not equal due to rounding error

Table G-3. Storage Potential for Unmineable Coal Seams Beneath Leasable Federal Land (High Estimate – Million Metric Tons)

RCSP REGION	TOTAL of Groups by RCSP	Group 1 Totals	Group 2 Totals	Group 3 Totals	Group 4 Totals	Group 5 Totals	Group 6 Totals	Group 7 Totals	Group 8 Totals	Group 9 Totals
		DOD	DOE	BLM	BOR	FWS	NPS	DOJ	Other Federal Agencies	FS and Other USDA
PCOR	1,663	0	0	645	0	0	0	0	0	1,018
MGSC	1	0	0	0	0	0	0	0	0	1
MRCSP	2	0	0	0	0	0	0	0	0	2
WESTCARB	0	0	0	0	0	0	0	0	0	0
SECARB	3,783	0	0	0	19	0	0	0	0	3,763
SWP	136	0	0	20	1	0	0	0	0	115
Big Sky	248	0	0	221	19	0	0	0	0	8
Northeast	0	0	0	0	0	0	0	0	0	0
Total*	5,833	0	0	886	39	0	0	0	0	4,908

* sum may not equal due to rounding error

Table G-4. Storage Potential for Unmineable Coal Seams Beneath Leasable Federal Land (Low Estimate – Million Metric Tons)

RCSP REGION	TOTAL of Groups by RCSP	Group 1 Totals	Group 2 Totals	Group 3 Totals	Group 4 Totals	Group 5 Totals	Group 6 Totals	Group 7 Totals	Group 8 Totals	Group 9 Totals
		DOD	DOE	BLM	BOR	FWS	NPS	DOJ	Other Federal Agencies	FS and Other USDA
PCOR	1,663	0	0	645	0	0	0	0	0	1,018
MGSC	1	0	0	0	0	0	0	0	0	1
MRCSP	1	0	0	0	0	0	0	0	0	1
WESTCAR B	0	0	0	0	0	0	0	0	0	0
SECARB	2,654	0	0	0	13	0	0	0	0	2,641
SWP	57	0	0	5	0	0	0	0	0	52
Big Sky	108	0	0	99	6	0	0	0	0	2
Northeast	0	0	0	0	0	0	0	0	0	0
Total*	4,483	0	0	750	20	0	0	0	0	3,713

* sum may not equal due to rounding error

End Notes

[1] U.S. Onshore storage capacity: low estimate 2,289 billion metric tons; high estimate 8,129 billion metric tons. 2008 Carbon Sequestration Atlas of the United States and Canada (2nd Edition). Found at: http://www.netl.doe.gov/technologies/carbon seq/refshelf/atlasII/index.html

[2] National Carbon Sequestration Database and Geographic Information System (NATCARB). http://www.natcarb.org/

[3] Ventyx Velocity Suite. http://www1.ventyx.com/velocity/vs-overview.asp

[4] Pipeline and Hazardous Materials Safety Administration (PHMSA) National Pipeline Mapping System (NPMS). http://www.npms.phmsa.dot.gov/

[5] Ibid, see footnote #1

[6] Big Sky Carbon Sequestration Partnership (Big Sky). http://www.bigskyco2.org/

[7] Midwest Geological Sequestration Consortium (MGSC). http://www.sequestration.org/

[8] Midwest Regional Carbon Sequestration Partnership (MRCSP). http://www.mrcsp.org/

[9] Plains CO2 Reduction Partnership (PCOR). http://www.undeerc.org/pcor/

[10] Southeast Regional Carbon Sequestration Partnership (SECARB). http://www.secarbon.org/

[11] Southwest Partnership CO2 Sequestration (SWP). http://www.southwestcarbonpartnership.org/

[12] West Coast Regional Carbon Sequestration Partnership (WESTCARB). http://www.westcarb.org/

[13] National Carbon Sequestration Database and Geographic Information System (NATCARB). http://www.natcarb.org/

[14] National Atlas: Federal Lands of the United States. http://www.nationalatlas.gov/mld/fedlanp.html

[15] Ibid.

[16] EPCA Phase III, *Inventory of Onshore Federal Oil and Natural Gas Resources and Restrictions to Their Development: Phase III Inventory*. Appendix 9. Prepared by the U.S. Departments of the Interior, Agriculture, and Energy. May 2008.

[17] Thompson, D. CRS Report for Congress: 96-161. Mining in National Parks and Wilderness Areas: Policy, Rules, Activity. 1996 Feb 12. http://www.ncseonline.org/nle/crsreports/mining/mine-6.cfm. Accessed June 26, 2008.

[18] Krista L. Edwards, Written Testimony for U.S. Senate Committee on Energy and Natural Resources: *Oversight Hearing on Construction and Operation of Carbon Dioxide Pipelines*. (Jan. 312008).

[19] *Carbon Sequestration Atlas of the United States and Canada*. (The Department of Energy's National Energy Technology Laboratory, March 2007).

[20] *Carbon Sequestration Atlas of the United States and Canada*. (The Department of Energy's National Energy Technology Laboratory, March 2007).

[21] *Carbon Sequestration Atlas of the United States and Canada*. (The Department of Energy's National Energy Technology Laboratory, March 2007), page 14.

[22] Development of a Probabilistic Assessment Methodology for Evaluation of Carbon Dioxide Storage, USGS Open File Report 2009-1035. Found at: http://pubs.usgs.gov/of/2009/1035/ofr2009-1035.pdf

[23] 30 U.S.C. §181 et seq.

[24] 30 U.S.C. §351-359.

[25] Found at: http://www.blm.gov/pgdata/etc/medialib/blm/es/og_ forms.Par.49482.File.dat/BLM_3100- 11%5B1%5D.pdf

[26] EPCA Phase II, *Scientific Inventory of Onshore Federal Lands' Oil and Gas Resources and the Extent and Nature of Restrictions or Impediments to Their Development: Phase II Cumulative Inventory*. Appendix 2, Glossary. Prepared by the U.S. Departments of the Interior, Agriculture, and Energy, November 2006

[27] 43 CFR §6301.3.

[28] D. Thompson, CRS Report for Congress: *Mining in National Parks and Wilderness Areas: Policy, Rules, Activity*. (Feb. 12 1996), 96-161 http://www.ncseonline.org/nle/crsreports/mining/mine-6.cfm. (Accessed June 26, 2008).

[29] EPCA Phase II, *Scientific Inventory of Onshore Federal Lands' Oil and Gas Resources and the Extent and Nature of Restrictions or Impediments to Their Development: Phase II Cumulative Inventory,* Appendix 4. Prepared by the U.S. Departments of the Interior, Agriculture, and Energy, November 2006.

[30] 42 U.S.C. §4321.

[31] 16 U.S.C. §470 *et seq.*

[32] 16 U.S.C. §1531 *et seq.*

[33] Report by the Federal Rights-of-Way Working Group, *Improving Rights-of-Way Management across Federal Lands: A Roadmap for Greater Broadband Deployment*, April 2004.

[34] 16 U.S.C. §1531 (a)(2)-(a)(3). Other examples of such laws include National Wildlife Refuge System Administration Act of 1966, as amended, (16 U.S.C. §§ 668dd-668ee); the Coastal Zone Management Act of 1972, as amended (16 U.S.C. §1451 *et seq.*); the Archaeological and Historic Preservation Act of 1974, as amended (16 U.S.C. §469 *et seq.*); Section 404 of the Federal Water Pollution Control Act (Clean Water Act), as amended (33 U.S.C. § 1344); Section 10 of the Rivers and Harbors Act of 1899, as amended (33 U.S.C. § 403); and the Comprehensive Environmental Response, Compensation, and Liability Act of 1980, as amended (42 U.S.C. § 9601 *et seq.*).

[35] National Petroleum Council (NPC), Balancing Natural Gas Policy, Fueling the Demands of a Growing Economy. Vol IV, Supply Task Group Report. September 2003. Table S6-2. Found at: http://www.npc.org/

[36] 16 U.S.C. § 1532.

[37] EPCA Phase II, *Scientific Inventory of Onshore Federal Lands' Oil and Gas Resources and the Extent and Nature of Restrictions or Impediments to Their Development: Phase II Cumulative Inventory*. Prepared by the U.S. Departments of the Interior, Agriculture, and Energy. November 2006.

[38] 30 U.S.C. § 185.

[39] Ibid.

[40] Adam Smith, *Regulatory Issues Controlling Carbon Capture and Storage*, MIT Masters Thesis, (June 2004): http://dspace.mit.edu/bitstream/handle/1721.1/27019/56748751.pdf?sequence=1

[41] 42 U.S.C. §4321.

[42] 16 U.S.C. §470 *et seq.*

[43] 16 U.S.C. §1531 *et seq.*

[44] Report by the Federal Rights-of-Way Working Group, *Improving Rights-of-Way Management Across Federal Lands: A Roadmap for Greater Broadband Deployment*, (April 2004).

[45] Report by the Federal Rights-of-Way Working Group, *Improving Rights-of-Way Management Across Federal Lands: A Roadmap for Greater Broadband Deployment*, (April 2004).

[46] "Public lands" are defined by the FLPMA as "any land and interest in land owned by the United States within the several States and administered by the Secretary of the Interior through the Bureau of Land Management, with regard to how the United States acquired

ownership, except – (1) lands located on the Outer Continental Shelf; and (2) lands held for the benefit of Indians, Aleuts, and Eskimos." 43 U.S.C.§1702.

[47] 43 U.S.C. §1765(a)(ii).

[48] 43 U.S.C. §1765(b)(i).

[49] 43 U.S.C. §1764.

[50] Report by the Federal Rights-of-Way Working Group, *Improving Rights-of-Way Management Across Federal Lands: A Roadmap for Greater Broadband Deployment*, April 2004.

[51] 43 U.S.C. §101(2).

[52] Hearing Before the Committee on Energy and Natural Resources United States Senate (110th Congress), *Regulatory Aspects of Carbon Capture, Transportation, and Sequestration*, (January 31, 2007).

[53] Railroad Commission of Texas, *Pipeline Eminent Domain and Condemnation*: FAQs, http://www.rrc.state.tx.us/eminentdomain.html (Accessed Sept. 29, 2008).

[54] 30 USC 185.

[55] Ibid.

[56] Adam Smith, *Regulatory Issues Controlling Carbon Capture and Storage*, MIT Masters Thesis, (June 2004): http://dspace.mit.edu/bitstream/handle/1721.1/27019/56748751.pdf?sequence=1

[57] *Mineral Leasing Act of 1920* as Amended, re-transcribed Aug. 9, 2007. http://www.blm.gov/pgdata/etc/medialib/blm/wo/Communications_Directorate/legislation.Par.23212.File.dat/mla_1 920_amendments1.pdf (Accessed Feb. 25, 2008).

[58] 42 USC 15926.

[59] http://corridoreis.anl.gov (Accessed Feb. 25, 2008).

[60] *Mineral Leasing Act of 1920* as Amended, re-transcribed Aug. 9, 2007. http://www.blm.gov/pgdata/etc/medialib/blm/wo/Communications_Directorate/legislation.Par.23212.File.dat/mla_1 920_amendments1.pdf (Accessed Feb. 25, 2008).

[61] *Alaska Natural Gas Development Authority Conditional Right-of-Way Lease*, Commissioner's Response to Public Comments and Final Decision, ADL 229297, (July 11, 2006). http://www.jpo.doi.gov/ANGDA/ANGDApercent20FINALpercent20REPORTS/ Commissioner's percent 20Final percent20Decision.pdf (Accessed August 26, 2008).

[62] National Telecommunications and Information Administration, *Rights-of-Way Laws by State*, Updated May 21, 2003, http://www.ntia.doc.gov/ntiahome/staterow/rowtableexcel.htm (Accessed August 26, 2008).

[63] 49 U.S.C. § 1671 *et seq*.

[64] 49 U.S.C. § 601.

[65] 49 C.F.R. § 190, 195-199.

[66] 42 USC 300f – j.

[67] Commonwealth of Pennsylvania, Department of Conservation and Natural Resources, *Report of the Carbon Management Advisory Group*, (May 2008).

[68] U.S. EPA UIC Program: *Classes of Wells*. http://www.epa.gov/safewater/uic/wells.html. (Accessed June 19, 2008).

[69] U.S. EPA, Office of Water 4604, *Technical Program Overview: Underground Injection Control Regulations*, EPA 816-R-02-025, (July 2001) www.epa.gov (Accessed June 10, 2008).

[70] *Using the Class V Experimental Technology Well Classification for Pilot Geologic Sequestration Projects – UIC Program Guidance (UICPG #83)*, (March 1, 2007). www.epa.gov (Accessed July 7, 2007).

[71] *Federal Requirements under the Underground Injection Control (UIC) Program for Carbon Dioxide (CO2) Geologic Sequestration (GS) Wells; Proposed Rule*, http://www.epa.gov/fedrgstr/EPA-WATER/2008/July/Day25/w16626.pdf (Accessed August 20, 2008).

[72] Mark A. de Figueiredo, *Property Interests and Liability of Geologic Carbon Dioxide Storage*, A Special Report to the MIT Carbon Sequestration Initiative, September 2005.

[73] Sarah Wade, *Addressing CCS Liability*, Presentation to Edison Institute, Washington D.C., (March 4, 2008).

[74] "Storage of Carbon Dioxide in Geologic Structures: A Legal and Regulatory Guide for States and Provinces," The Interstate Oil and Gas Compact Commission, September 25, 2007.

[75] Mark A. de Figueiredo, *Property Interests and Liability of Geologic Carbon Dioxide Storage*, A Special Report to the MIT Carbon Sequestration Initiative, (September 2005). Found at: http://sequestration.mit.edu/pdf/deFigueiredo_Property_Interests.pdf

[76] Title VII, Section 702, Energy Independence and Security Act of 2007. Found at: http://frwebgate.access.gpo.gov/cgi-bin/getdoc.cgi?dbname=110_cong_public_laws&docid=f:publ140.110

[77] Ibid.

[78] http://www.govtrack.us/congress/bill.xpd?bill=s110-2191 (Accessed June 17, 2008).

[79] Carbon Capture and Storage Technology Act of 2007. Found at: http://thomas.loc.gov/cgi-bin/query/z?c110:S.2323:

[80] Carbon Dioxide Pipeline Study Act of 2007. Found at: http://thomas.loc.gov/cgi-bin/query/z?c110:S.2144:

[81] Chapter 173-218 WAC "Underground Injection Control Program," www.ecy.wa.gov, Accessed July 14, 2008.

[82] GovTrack.us. S. 2323--110th Congress (2007): *Carbon Capture and Storage Technology Act of 2007*, http://www.govtrack.us/congress/bill.xpd?bill=s110-2323 (Accessed Sep 19, 2008).

[83] GovTrack.us. S. 3036--110th Congress (2008): *Lieberman-Warner Climate Security Act of 2008*, http://www.govtrack.us/congress/bill.xpd?bill=s110-3036&tab=related (Accessed Sep 19, 2008).

[84] The Interstate Oil and Gas Compact Commission, *Storage of Carbon Dioxide in Geologic Structures: A Legal and Regulatory Guide for States and Provinces*, (September 25, 2007).

[85] National Petroleum Council (NPC), Balancing Natural Gas Policy, Fueling the Demands of a Growing Economy. Vol IV, Supply Task Group Report. September 2003. Table S6-2. Found at: http://www.npc.org/

[86] Carbon Monitoring for Action (CARMA). http://carma.org/

[87] Ventyx Velocity Suite. http://www1.ventyx.com/velocity/vs-overview.asp

[88] Plains CO2 Reduction Partnership (PCOR). http://www.undeerc.org/pcor/

[89] Midwest Regional Carbon Sequestration Partnership (MRCSP). http://www.mrcsp.org/

[90] National Carbon Sequestration Database and Geographic Information System (NATCARB). http://www.natcarb.org/

[91] Ventyx Velocity Suite. http://www1.ventyx.com/velocity/vs-overview.asp

[92] Pipeline and Hazardous Materials Safety Administration (PHMSA) National Pipeline Mapping System (NPMS). http://www.npms.phmsa.dot.gov/

In: Geologic Carbon Dioxide Storage … ISBN: 978-1-63483-560-2
Editor: Theresa Wise

Chapter 2

NATIONAL ASSESSMENT OF GEOLOGIC CARBON DIOXIDE STORAGE RESOURCES—ALLOCATIONS OF ASSESSED AREAS TO FEDERAL LANDS*

Marc L. Buursink, Steven M. Cahan and Peter D. Warwick

CONVERSION FACTORS

Multiply	By	To obtain
	Length	
foot (ft)	0.3048	meter (m)
mile (mi)	1.609	kilometer (km)
nautical mile (nmi)	1.852	kilometer (km)
	Area	
acre	4,047	square meter
square mile	2.590	square kilometer
	Mass	
kilogram (kg)	2.205	pound, avoirdupois (lb)
megagram (Mg)=1 metric ton (t) (1,000 kg)	1.102	ton, short (2,000 lb)
billion metric tons = gigaton (Gt)	1.102	billion short tons

* This is an edited, reformatted and augmented version of U.S. Geological Survey Scientific Investigations Report 2015–5021, approved February 2015.

Datum

Horizontal coordinate information is referenced to the North American Datum of 1983 (NAD 83).

Abstract

Following the geologic basin-scale assessment of technically accessible carbon dioxide storage resources in onshore areas and State waters of the United States, the U.S. Geological Survey estimated that an area of about 130 million acres (or about 200,000 square miles) of Federal lands overlies these storage resources. Consequently, about 18 percent of the assessed area associated with storage resources is allocated to Federal land management. Assessed areas are allocated to four other general land-ownership categories as follows: State lands about 4.5 percent, Tribal lands about 2.4 percent, private and other lands about 72 percent, and offshore areas about 2.6 percent.

Introduction

In 2012, the U.S. Geological Survey (USGS) completed an assessment of the technically accessible geologic storage resources (TA_{SR}) for carbon dioxide (CO_2) for 36 assessed areas (within one or more sedimentary basins) in 8 regions in the onshore areas and State waters of the United States (figure 1) (U.S. Geological Survey Geologic Carbon Dioxide Storage Resources Assessment Team, 2013a,b,c). The TA_{SR} is an estimate of the subsurface geologic storage resources that may be available for CO_2 injection and is based on current geologic and hydrologic knowledge of the subsurface and on assumptions about current engineering practices (Society of Petroleum Engineers, 2014). By using a geology-based probabilistic assessment methodology, the USGS assessment team members obtained a mean TA_{SR} estimate of about 3,000 gigatons (Gt) under U.S. onshore areas and State waters. The estimate of the TA_{SR} includes buoyant trapping storage resources (B_{SR}), where CO_2 can be trapped in structural or stratigraphic closures, and residual trapping storage resources (R_{SR}), where CO_2 can be held in place by capillary pore pressures in areas outside of buoyant traps. The mean national B_{SR} is about 44 Gt, whereas the R_{SR} is divided among three injectivity classes, which are based on reservoir permeability. A large portion of the TA_{SR} is located under Federal lands, and identifying this fraction of the assessed area is critical to future management decisions.

The Energy Independence and Security Act of 2007 (Public Law 110–140) directed the USGS to conduct a national assessment of geologic storage resources for CO_2 in consultation with the U.S. Environmental Protection Agency (EPA), the U.S. Department of Energy (DOE), and State geological surveys. The USGS subsequently developed and updated a probabilistic methodology to estimate storage resources in U.S. sedimentary basins (Burruss and others, 2009; Brennan and others, 2010; Blondes and others, 2013). The national USGS assessment results were summarized by the U.S. Geological Survey Geologic Carbon Dioxide Storage Resources Assessment Team (2013c) and were detailed in companion reports (U.S. Geological Survey Geologic Carbon Dioxide Storage Resources Assessment Team, 2013a,b).

STORAGE RESOURCES

The goal of the national assessment by the USGS was to prepare an estimate of subsurface storage capacity on a regional scale; the results are not intended for use in the evaluation of specific sites for potential CO_2 storage. The assessment was a geology-based examination of major sedimentary basins in the onshore areas and State waters of the United States that contain one or more storage assessment units (SAUs) defined according to geologic and hydrologic characteristics required by the USGS assessment methodology. Although geologic storage of CO_2 may be possible in some areas not assessed by the USGS, the SAUs identified in the assessment represent those areas within sedimentary basins that met the assessment criteria. The SAU is a mappable volume of rock that consists of a porous reservoir and a bounding regional sealing formation. A geologic description of each sedimentary basin and the SAUs therein was prepared during the assessment; assessment input parameters for the SAUs in several basins have been released in Warwick and Corum (2012).

Sedimentary rocks of deep formations with saline groundwater and of reservoirs containing existing oil and gas fields were evaluated. Ultimately, 36 sedimentary basins, or combined basin study areas, within 8 regions of the United States were assessed for storage resources (U.S. Geological Survey Geologic Carbon Dioxide Storage Resources Assessment Team, 2013a,b) (figure 1). Within the assessed basins, 202 SAUs were identified as having good storage potential because of the presence of a robust regional seal, adequate reservoir rock, and sufficient areas containing saline formation waters. Regions with the largest TA_{SR} are the Coastal Plains Region (mean estimate of 1,900 Gt, of which about 1,800 Gt, or 95 percent of the resources, is in the U.S. Gulf Coast study area), the Alaska Region (mean estimate of 270 Gt, where the resources are almost entirely in the Alaska North Slope), and the Rocky Mountains and Northern Great Plains Region (mean estimate of 270 Gt) (table 1; figure 1) (U.S. Geological Survey Geologic Carbon Dioxide Storage Resources Assessment Team, 2013a,b). No SAUs were identified in Hawaii because geologic conditions were unlikely to meet the USGS methodology requirements.

Table 1. U.S. regions and basins assessed in 2012 for geologic carbon dioxide storage, merged storage assessment unit (SAU) footprint areas, and estimated areas for lands overlying storage resources and allocated to (*A*) general land-ownership categories and (*B*) Federal land-management categories

[The 2012 assessment data are from the U.S. Geological Survey Geologic Carbon Dioxide Storage Resources Assessment Team (2013a). Resources underlying federally managed offshore areas were not assessed. Allocations were calculated from unrounded assessed areas but are reported to two significant figures. The assessed areas may underlie (table 1*A*) general land-ownership categories including Tribal lands or (table 1*B*) Federal lands, including in order of prevalence, lands managed by the Bureau of Land Management (BLM), U.S. Department of Agriculture Forest Service (FS), U.S. Department of Defense (DOD), U.S. Fish and Wildlife Service (USFWS), National Park Service (NPS), and Bureau of Reclamation (BOR). For completeness, Tennessee Valley Authority lands are shown in figure 1; because they do not overlie assessed areas, they are omitted from this table.—, no allocation occurrence]

A. General land-ownership categories						
Assessed area (assessment basin code)	Merged SAU footprint (most likely), in acres	Allocations (most likely) to general land-ownership categories, in acres				
		Federal lands	State lands	Tribal lands	Private and other lands	Offshore areas
Alaska Region						
Alaska North Slope (C5001)	41,000,000	25,000,000	9,200,000	4,000,000	23,000	2,400,000
Kandik Basin (C5002)	1,200,000	940,000	23,000	230,000	—	—
Total	42,000,000	26,000,000	9,200,000	4,200,000	23,000	2,400,000

A. General land-ownership categories						
Assessed area (assessment basin code)	Merged SAU footprint (most likely), in acres	Allocations (most likely) to general land-ownership categories, in acres				
		Federal lands	State lands	Tribal lands	Private and other lands	Offshore areas
Pacific Northwest Region						
Columbia Basin of Oregon, Washington, and Idaho (C5005)	16,000,000	2,300,000	380,000	660,000	13,000,000	—
Western Oregon and Washington Basins (C5004)	2,100,000	47,000	88,000	11,000	1,800,000	230,000
Total	**18,000,000**	**2,400,000**	**470,000**	**670,000**	**15,000,000**	**230,000**
California Region						
Central California Coast Basins (C5011)	570,000	77,000	740	—	490,000	—
Los Angeles Basin (C5014)	270,000	3,100	250	—	270,000	3.3
Sacramento Basin (C5009)	1,800,000	19,000	63,000	—	1,700,000	—
San Joaquin Basin (C5010)	5,500,000	180,000	58,000	40	5,300,000	—
Ventura Basin (C5013)	580,000	18,000	4,000	—	370,000	190,000
Total	**8,700,000**	**300,000**	**130,000**	**40**	**8,100,000**	**190,000**
Rocky Mountains and Northern Great Plains Region						
Bighorn Basin (C5034)	4,300,000	2,600,000	280,000	6,500	1,400,000	—
Denver Basin (C5039)	24,000,000	460,000	1,400,000	—	23,000,000	—
Eastern Great Basin (C5019)	42,000,000	35,000,000	900,000	230,000	5,600,000	—
Greater Green River Basin (C5037)	15,000,000	9,300,000	640,000	—	4,800,000	—
Hanna, Laramie, and Shirley Basins (C5030)	2,600,000	860,000	180,000	—	1,600,000	—
Paradox Basin (C5021)	6,100,000	3,600,000	350,000	830,000	1,300,000	—
Powder River Basin (C5033)	17,000,000	2,500,000	1,100,000	1,000,000	12,000,000	—
Raton Basin (C5041)	1,300,000	66,000	36,000	—	1,200,000	—
San Juan Basin (C5022)	6,500,000	2,100,000	280,000	3,000,000	1,100,000	—
Uinta and Piceance Basins (C5020)	12,000,000	7,200,000	790,000	850,000	3,000,000	—
Williston Basin (C5031)	62,000,000	7,500,000	2,500,000	4,600,000	48,000,000	—
Wind River Basin (C5035)	3,600,000	1,400,000	200,000	740,000	1,200,000	—
Wyoming-Idaho-Utah Thrust Belt (C5036)	7,000,000	3,500,000	430,000	110,000	2,900,000	—
Total	**200,000,000**	**76,000,000**	**9,200,000**	**11,000,000**	**110,000,000**	—
Western Mid-Continent Region						
Anadarko and Southern Oklahoma Basins (C5058)	35,000,000	770,000	480,000	—	33,000,000	—
Arkoma Basin (C5062)	7,900,000	470,000	61,000	—	7,300,000	—
Bend Arch and Fort Worth Basin (C5045)	19,000,000	210,000	12,000	—	18,000,000	—
Kansas Basins (C5056)	2,800,000	29,000	—	—	2,800,000	—
Palo Duro Basin (C5043)	8,600,000	22,000	120,000	—	8,500,000	—
Permian Basin (C5044)	19,000,000	1,500,000	840,000	—	16,000,000	—
Total	**91,000,000**	**3,000,000**	**1,500,000**	—	**87,000,000**	—
Eastern Mid-Continent Region						
Appalachian Basin (C5067)	60,000,000	3,000,000	2,900,000	62,000	52,000,000	2,400,000
Black Warrior Basin (C5065)	1,900,000	40,000	3,900	—	1,900,000	—
Illinois Basin (C5064)	42,000,000	1,800,000	—	—	40,000,000	—
Michigan Basin (C5063)	30,000,000	1,900,000	2,600,000	130,000	19,000,000	6,600,000
Total	**130,000,000**	**6,700,000**	**5,600,000**	**200,000**	**110,000,000**	**9,000,000**

Table 1. (Continued)

A. General land-ownership categories

Assessed area (assessment basin code)	Merged SAU footprint (most likely), in acres	Allocations (most likely) to general land-ownership categories, in acres				
		Federal lands	State lands	Tribal lands	Private and other lands	Offshore areas
Coastal Plains Region						
Atlantic Coastal Plain (C5070)	14,000,000	990,000	270,000	—	11,000,000	2,000,000
South Florida Basin (C5050)	22,000,000	3,300,000	2,300,000	190,000	12,000,000	4,400,000
U.S. Gulf Coast (C5047 and C5049)	150,000,000	7,800,000	2,100,000	19,000	140,000,000	4,700
Total	**190,000,000**	**12,000,000**	**4,600,000**	**210,000**	**160,000,000**	**6,400,000**
Eastern Mesozoic Rift Basins Region						
Eastern Mesozoic Rift Basins (C5068)	1,300,000	13,000	30,000	—	1,300,000	—
Total	**1,300,000**	**13,000**	**30,000**	—	**1,300,000**	—
Grand total	**690,000,000**	**130,000,000**	**31,000,000**	**17,000,000**	**490,000,000**	**18,000,000**

B. Federal land-management categories

Assessed area (assessment basin code)	Allocations (most likely) to Federal land-management categories, in acres						
	BLM	FS	DOD	USFWS	NPS	Other	BOR
Alaska Region							
Alaska North Slope (C5001)	25,000,000	—	45	420,000	120,000	—	—
Kandik Basin (C5002)	560,000	—	—	—	380,000	—	—
Total	**25,000,000**	—	**45**	**420,000**	**500,000**	—	—
Pacific Northwest Region							
Columbia Basin of Oregon, Washington, and Idaho (C5005)	360,000	970,000	410,000	57,000	87	380,000	160,000
Western Oregon and Washington Basins (C5004)	42,000	—	1,300	2,800	—	130	—
Total	**400,000**	**970,000**	**410,000**	**60,000**	**87**	**380,000**	**160,000**
California Region							
Central California Coast Basins (C5011)	16,000	22,000	39,000	—	—	—	—
Los Angeles Basin (C5014)	—	—	2,100	950	—	—	—
Sacramento Basin (C5009)	1,700	—	9,900	3,600	—	3,700	98
San Joaquin Basin (C5010)	70,000	—	81,000	30,000	—	—	—
Ventura Basin (C5013)	2,800	12,000	1,500	1,900	—	—	—
Total	**90,000**	**34,000**	**130,000**	**36,000**	—	**3,700**	**98**
Rocky Mountains and Northern Great Plains Region							
Bighorn Basin (C5034)	2,300,000	150,000	350	—	—	—	78,000
Denver Basin (C5039)	29,000	300,000	62,000	49,000	7,000	3,900	1,800
Eastern Great Basin (C5019)	27,000,000	2,700,000	3,300,000	810,000	340,000	570,000	30,000
Greater Green River Basin (C5037)	8,100,000	920,000	—	23,000	—	—	210,000
Hanna, Laramie, and Shirley Basins (C5030)	820,000	17,000	—	1,200	—	—	23,000
Paradox Basin (C5021)	2,500,000	630,000	1,600	—	440,000	—	89,000
Powder River Basin (C5033)	1,700,000	830,000	13,000	—	760	—	1,200
Raton Basin (C5041)	29,000	37,000	—	—	—	—	—
San Juan Basin (C5022)	1,700,000	360,000	—	—	34,000	—	31,000
Uinta and Piceance Basins (C5020)	4,700,000	2,400,000	54,000	9,000	18,000	—	—
Williston Basin (C5031)	2,300,000	3,700,000	610,000	710,000	73,000	91,000	130,000
Wind River Basin (C5035)	1,300,000	19,000	1,200	—	—	—	120,000
Wyoming-Idaho-Utah Thrust Belt (C5036)	1,200,000	2,200,000	—	6,300	8,300	40,000	9,700
Total	**54,000,000**	**14,000,000**	**4,100,000**	**1,600,000**	**920,000**	**710,000**	**720,000**

B. Federal land-management categories							
Assessed area (assessment basin code)	Allocations (most likely) to Federal land-management categories, in acres						
	BLM	FS	DOD	USFWS	NPS	Other	BOR
Western Mid-Continent Region							
Anadarko and Southern Oklahoma Basins (C5058)	11,000	560,000	150,000	19,000	6,400	—	21,000
Arkoma Basin (C5062)	—	150,000	280,000	36,000	43	5,900	—
Bend Arch and Fort Worth Basin (C5045)	—	120,000	96,000	—	—	—	—
Kansas Basins (C5056)	—	—	29,000	—	—	—	—
Palo Duro Basin (C5043)	—	—	3,500	18,000	—	—	—
Permian Basin (C5044)	1,200,000	120,000	—	—	120,000	10,000	410
Total	**1,200,000**	**950,000**	**560,000**	**73,000**	**130,000**	**16,000**	**21,000**
Eastern Mid-Continent Region							
Appalachian Basin (C5067)	—	2,700,000	41,000	30,000	220,000	310	—
Black Warrior Basin (C5065)	—	26,000	11,000	—	2,700	—	—
Illinois Basin (C5064)	—	1,300,000	400,000	110,000	720	1,000	—
Michigan Basin (C5063)	—	1,800,000	16,000	9,900	59,000	—	—
Total	—	**5,800,000**	**470,000**	**150,000**	**280,000**	**1,300**	—
Coastal Plains Region							
Atlantic Coastal Plain (C5070)	—	32,000	360,000	520,000	75,000	1,500	—
South Florida Basin (C5050)	—	200	140,000	640,000	2,500,000	—	—
U.S. Gulf Coast (C5047 and C5049)	—	5,000,000	1,100,000	1,300,000	320,000	67,000	17,000
Total	—	**5,000,000**	**1,600,000**	**2,500,000**	**2,900,000**	**68,000**	**17,000**
Eastern Mesozoic Rift Basins Region							
Eastern Mesozoic Rift Basins (C5068)	—	—	1,800	—	11,000	—	—
Total	—	—	**1,800**	—	**11,000**	—	—
Grand total	**80,000,000**	**27,000,000**	**7,200,000**	**4,800,000**	**4,700,000**	**1,200,000**	**920,000**

Figure 1 - Maps of the conterminous United States and Alaska showing selected landownership and land-management categories, along with the 8 regions (separated by bold dashed lines) and 36 areas (pattern) assessed in 2012 for geologic carbon dioxide storage (U.S. Geological Survey Geologic Carbon Dioxide Storage Resources Assessment Team, 2013b). Resources underlying federally managed offshore areas were not assessed, and Hawaii was considered unlikely to have significant storage resources. Spatial data came from the Bureau of Land Management's State-specific spatial resources available at http://www.blm.gov and from the National Atlas of the United States (Minerals Management Service, 2005; U.S. Geological Survey, 2005a,b). *A*, Federal lands and lands from selected general land-ownership categories overlying assessed areas listed in order of prevalence, including Bureau of Land Management (BLM), U.S. Department of Agriculture Forest Service (FS), Tribal, Department of Defense (DOD), U.S. Fish and Wildlife Service (USFWS), National Park Service (NPS), Bureau of Reclamation (BOR), and Tennessee Valley Authority (TVA) lands. Assessed areas are numbered with the last two digits of the basin codes used in table 1. *B*, BLM

lands overlying assessed areas. *C*, FS lands overlying assessed areas. *D*, Tribal lands overlying assessed areas.

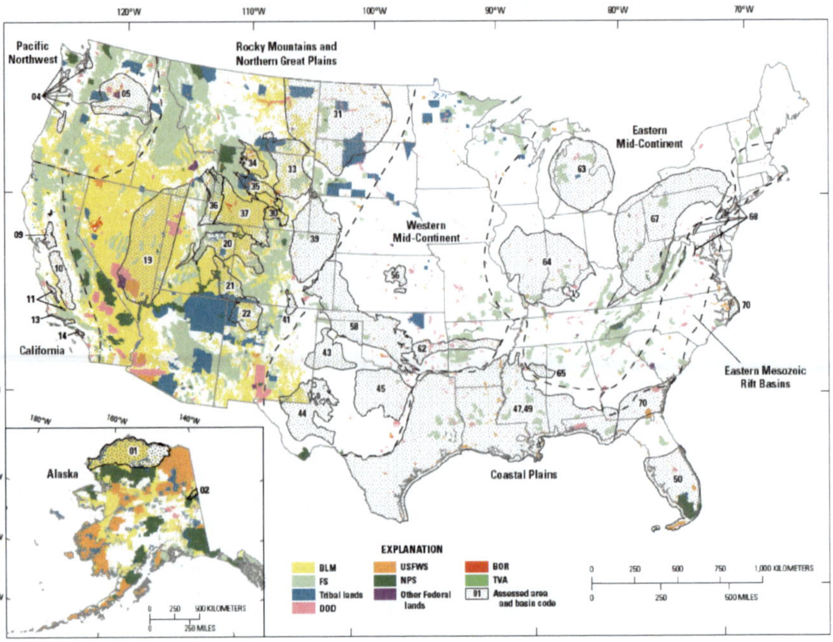

Figure 1A. Federal lands and lands from selected general land-ownership categories overlying assessed areas listed in order of prevalence (BLM, FS, Tribal, DOD, USFWS, NPS, other, BOR, TVA).

Storage resources were estimated without consideration of economic viability or of accessibility due to land-management or regulatory restrictions. Thus, if storage of CO_2 within a formation was feasible according to current knowledge of the subsurface and engineering practices, it was considered for the assessment. In the 2012 assessment, allocations to individual States and general land-ownership categories were tabulated for SAUs (U.S. Geological Survey Geologic Carbon Dioxide Storage Resources Assessment Team, 2013a).

In this report, potential resource accessibility related to general land ownership and Federal land management is now considered. The assessed area is used when allocating the storage resource because the basin geology may be too variable to allocate the TA_{SR} to general and Federal lands.

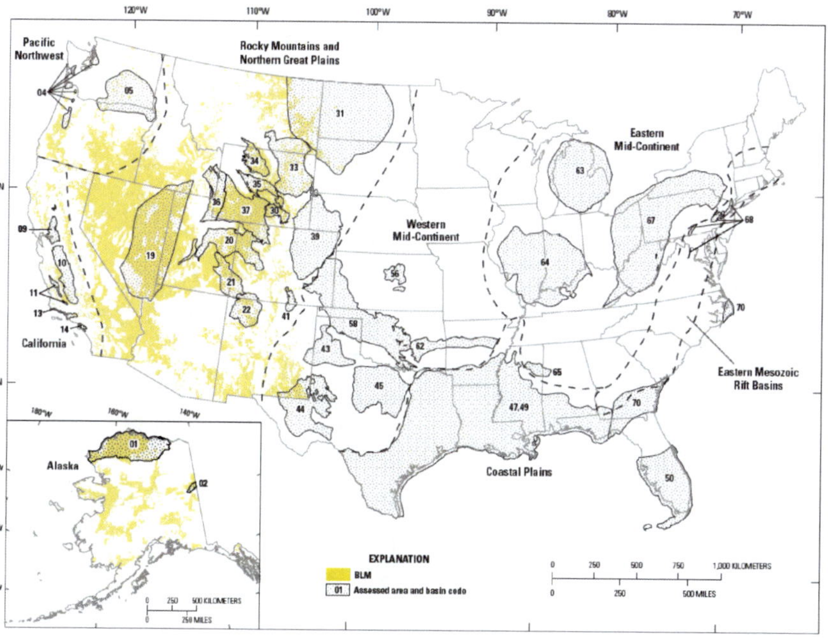

Figure 1B. Bureau of Land Management lands overlying assessed areas.

Furthermore, the individual agencies managing Federal lands overlying the assessed storage areas are now considered. The DOE published a complementary summary report of geologic CO_2 storage resources estimated by multiple public-private partnerships studies, and the report also accounts for Federal lands leasing (Dolence and others, 2009).

Lands Description

The five general land-ownership categories used in this report are Federal lands, State lands, Tribal lands, private and other lands, and offshore areas. The Federal lands considered in this report include lands where surface or subsurface rights are owned or managed by Federal agencies alone or in cooperation with State or private interests. Typically, officials at individual Federal sites have estimated the percentage of land on which the Federal Government owns the mineral rights. Also considered are lands held in trust by the Federal Government. In the conterminous United States and Alaska, Federal lands make up about 32 percent of the onshore areas and State waters.

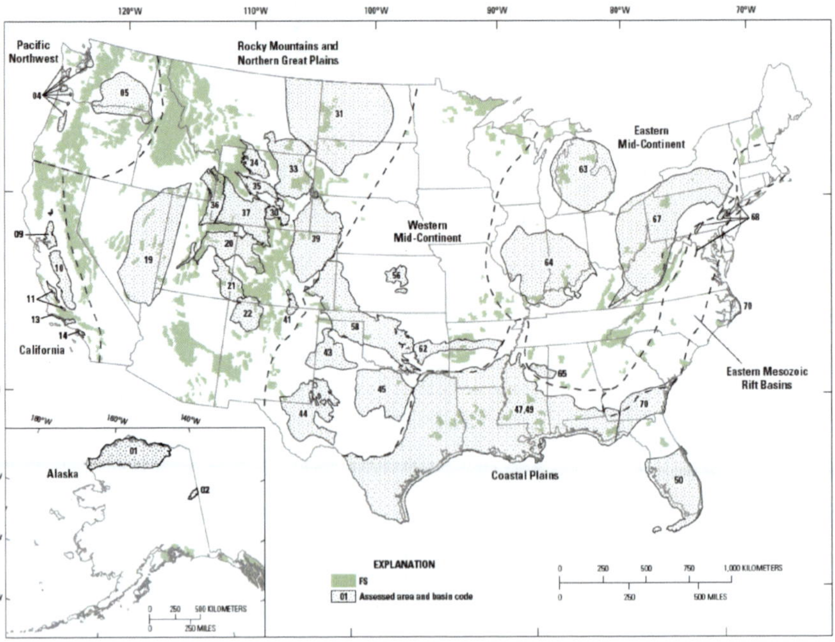

Figure 1C. U.S. Department of Agriculture Forest Service lands overlying assessed areas.

About 2.3 percent of the total U.S. area consists of Native American and Alaska Native lands (collectively referred to as Tribal lands; see the glossary), and they are considered in this report; these lands were not all included in the prior allocation by the U.S. Geological Survey Geologic Carbon Dioxide Storage Resources Assessment Team (2013a). Ultimately, land owners, managers, planners, and storage resource developers may use this report to make decisions about the multiple uses of Federal lands and the scope of future TA_{SR} development. State lands may have similar access and usage issues as Federal lands.

Agencies managing Federal lands within the assessment study areas include multiple U.S. Department of the Interior agencies, the U.S. Department of Agriculture Forest Service (FS), and the U.S. Department of Defense (DOD) (figure 1). The Interior agencies include the Bureau of Land Management (BLM), the Bureau of Reclamation (BOR), the U.S. Fish and Wildlife Service (USFWS), and the National Park Service (NPS). The U.S. Government does not have title to all the lands within the Federal tracts shown in figure 1.

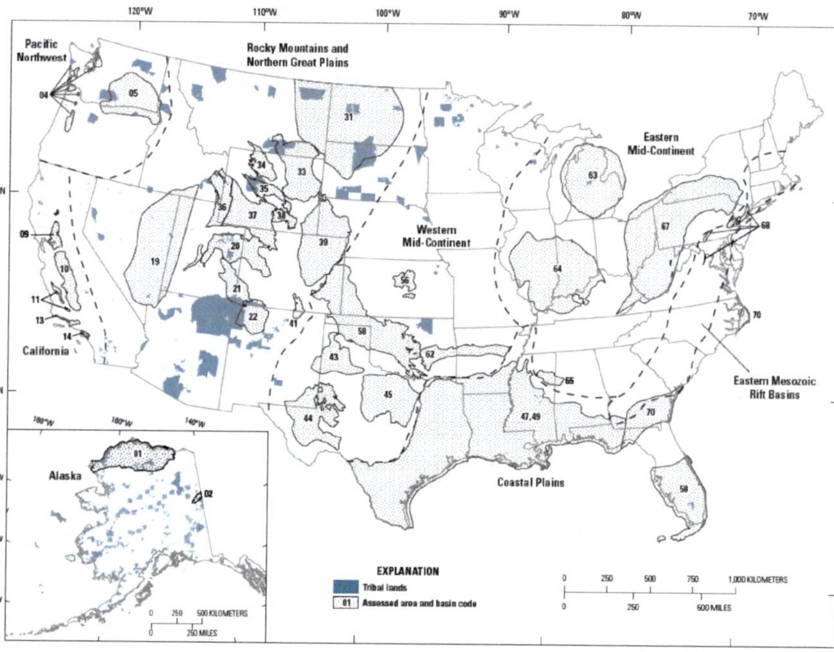

Figure 1D. Tribal lands overlying assessed areas.

In some areas, such as in legislatively proclaimed administrative boundaries of national forests, Federal and private surface ownerships are intermixed, as in Wyoming (figure 1). In other areas, such as small national wildlife refuges, the surface may be managed through easements rather than outright Federal ownership. Most of the land to which the U.S. Government holds title within the U.S. Gulf Coast study area was acquired from other owners (States and private) and may be subject to different public land laws and regulations when compared to land retained from the original public domain (Karlsen and others, 2002). Federal lands in the Eastern United States include national forests, U.S. military properties, national parks, the Tennessee Valley Authority (TVA) land, water bodies, other scattered recreational areas, and historical sites (figure 1). Federal lands in the Western United States include large BLM grazing and conservation areas, dammed water bodies, and extensive national forests.

Offshore areas are defined as State waters, or those submerged areas between the established baseline for the coast and the federally owned offshore areas. State waters extend out to 3 geographic (nautical) miles from the established baseline for the coast; beyond this limit, Federal jurisdiction begins and extends to an outer limit of 200 nautical miles. However, in special cases because of territorial claims existing at the dates of statehood, "Texas and the Gulf Coast of Florida have a proprietary interest in a submerged belt of land, nine geographic miles wide, extending seaward along the coast" (Thormahlen, 1999, p. 1–2). Resource assessments in federally managed areas beyond State waters (see the glossary) are typically done by the Bureau of Ocean Energy Management of the U.S. Department of the Interior.

Assessed Areas

An assessed area is the merged footprint for all the SAUs in a sedimentary basin or combined basin study area; the SAUs are at least 3,000 feet below the surface. The most likely estimate of assessed area represents the areal extent of the geologic storage resources and does not indicate uniform TA_{SR} distribution. For example, assessed areas may have overlapping subsurface SAUs, and the thickness of an SAU may vary within a basin. Specific information on each SAU was detailed in previous reports (Warwick and Corum, 2012; U.S. Geological Survey Geologic Carbon Dioxide Storage Resources Assessment Team, 2013a,b). Figure 1*A* shows assessed areas on a map of selected land-ownership and land-management categories, including Federal lands. For clarity, figures 1*B*, *C*, and *D* show the distribution of BLM, FS, and Tribal lands separately relative to the assessed areas.

Table 1*A* lists the most likely acreage in the merged SAU footprint for each assessed area and includes the allocations of Federal, State, Tribal, and other lands overlying each assessed area. Table 1*B* lists the most likely acreage of Federal lands overlying assessed areas and managed by different agencies. As shown in table 1*A*, the three assessed basins having the largest areas of overlying Federal lands are the Eastern Great Basin, the Alaska North Slope, and the Greater Green River Basin. Note that in the Eastern Great Basin, Federal lands overlie a large assessed area, but the basin contains only one quantitatively assessed SAU, which has a relatively low TA_{SR} (U.S. Geological Survey Geologic Carbon Dioxide Storage Resources Assessment Team, 2013b, table 3, p. 30 –31).

Figure 2*A* shows the most likely allocations in acres for five general land-ownership categories overlying assessed areas, whereas figure 2*B* shows the most likely allocations in acres for Federal land-management categories overlying assessed areas.

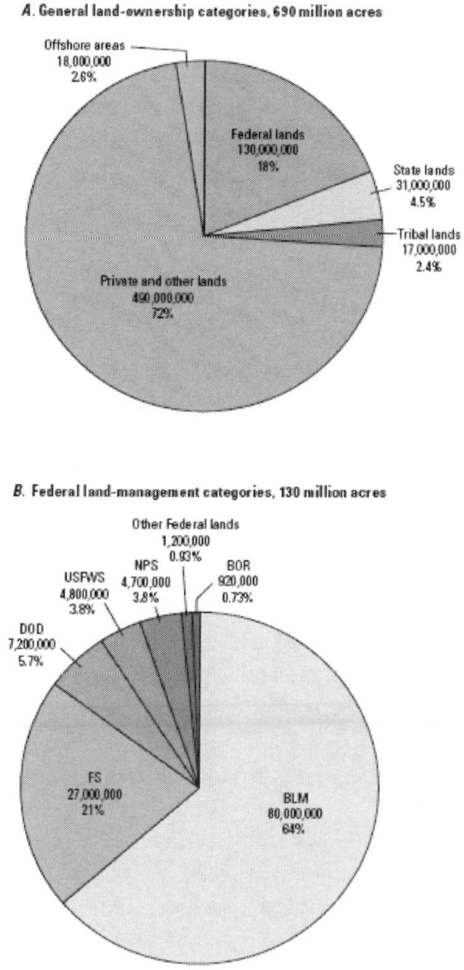

Figure 2. Pie charts showing land categories, along with most likely acreages and percentages for allocations of land overlying areas assessed in 2012 for geologic carbon dioxide storage (U.S. Geological Survey Geologic Carbon Dioxide Storage Resources Assessment Team, 2013a). A, Allocations to general land-ownership categories; B, Allocations to Federal land-management categories. Resources underlying federally managed offshore areas were not assessed. Allocation areas, in acres and percentages, were calculated from unrounded assessed areas but are reported to two significant figures. Terms: BLM, Bureau of Land Management; BOR, Bureau of Reclamation; DOD, Department of Defense; FS, U.S. Department of Agriculture Forest Service; NPS, National Park Service; USFWS, U.S. Fish and Wildlife Service; %, percent.

Federal lands overlie about 18 percent of the assessed acreage, whereas State lands overlie about 4.5 percent; Tribal lands, about 2.4 percent; private and other lands, about 72 percent; and offshore areas, about 2.6 percent. (Allocation areas, in acres and percentages, were calculated from unrounded assessed areas but are reported to two significant figures.) Of the assessed area overlain by Federal lands, 64 percent is managed by the BLM, 21 percent is managed by the FS, 5.7 percent is managed by the DOD, 3.8 percent is managed by the USFWS, and 3.8 percent is managed by the NPS; other agencies manage the remaining 1.7 percent (figure 2*B*).

Figures 1*B* and 2*B* and tables 1*B* and 2*B* show that the largest assessed area underlying Federal lands is allocated to lands administered by the BLM, particularly in the Alaska North Slope of the Alaska Region and in the Greater Green River Basin of the Rocky Mountains and Northern Great Plains Region. Figures 1*C* and 2*B* and tables 1*B* and 2*C* show that the second largest assessed area underlying Federal lands is allocated to lands administered by the FS, though no regions dominate the allocations.

Alternatively to assessed areas being allocated to land categories, the fraction of selected land categories overlying assessed areas may be considered. The fraction of all Federal lands overlying assessed areas is 19 percent, whereas the fraction of all BLM lands overlying assessed areas is 31 percent, the fraction of all FS lands overlying assessed areas is 13 percent, and the fraction of all Tribal lands overlying assessed areas is 18 percent.

Figure 3*A* shows the top 10 States (and other remaining States) having the largest estimated acreages of Federal lands overlying assessed areas, whereas figures 3*B* through 3*D* show the top 10 States having the largest estimated acreages of BLM, FS, and Tribal lands, respectively, overlying assessed areas. In general, the western States dominate these rankings. The States that have the most Federal lands overlying assessed areas are Alaska and Nevada, and each has about 26 million acres (or about 41,000 and 40,000 square miles, respectively) of mostly BLM lands. The States that have the least Federal lands overlying assessed areas are Missouri, with about 72 acres (or about 0.11 square mile) allocated to NPS lands, and Maryland, with about 11 acres (or about 0.017 square mile) allocated to DOD lands. Nearly all of the Federal lands in Louisiana overlie assessed areas (figure 1*A*). Twelve States have no assessed area and thus no allocations to general and Federal categories.

Nationwide, about 31 million acres (or about 48,000 square miles) of State lands overlie assessed areas.

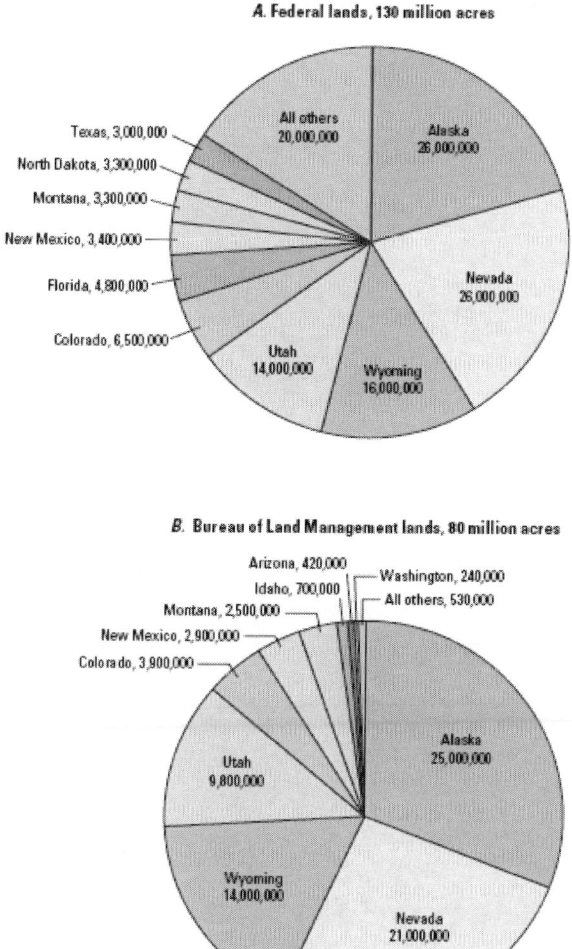

Figure 3. Pie charts showing top 10 States (and other remaining States) with the most likely estimates of allocations to four categories of land overlying areas assessed in 2012 for geologic carbon dioxide storage (U.S. Geological Survey Geologic Carbon Dioxide Storage Resources Assessment Team, 2013a). A, Allocations to Federal lands; B, Allocations to Bureau of Land Management lands; C, Allocations to U.S. Department of Agriculture Forest Service lands; and D, Allocations to Tribal lands. Resources underlying federally managed offshore areas were not assessed. Allocation areas, in acres, were calculated from unrounded assessed areas but are reported to two significant figures.

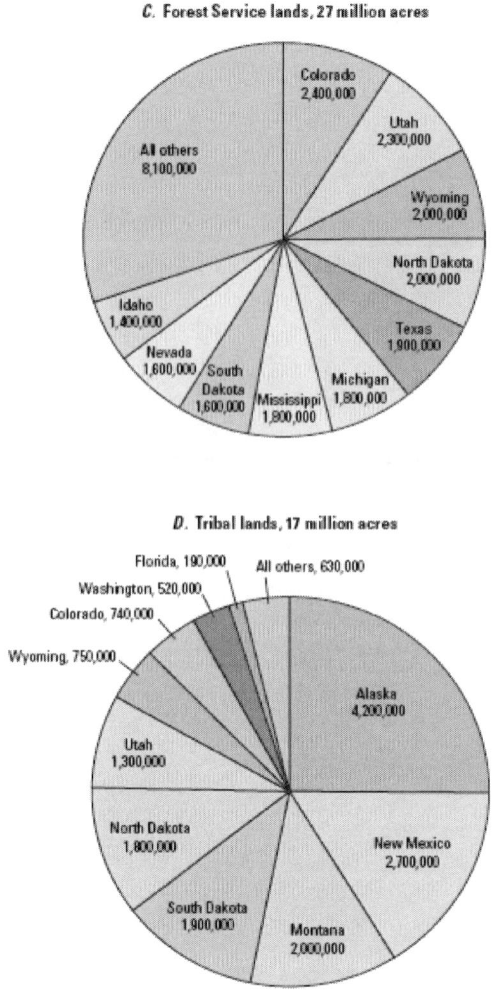

Figure 3. (Continued) Pie charts showing top 10 States (and other remaining States) with the most likely estimates of allocations to four categories of land overlying areas assessed in 2012 for geologic carbon dioxide storage (U.S. Geological Survey Geologic Carbon Dioxide Storage Resources Assessment Team, 2013a). A, Allocations to Federal lands; B, Allocations to Bureau of Land Management lands; C, Allocations to U.S. Department of Agriculture Forest Service lands; and D, Allocations to Tribal lands. Resources underlying federally managed offshore areas were not assessed. Allocation areas, in acres, were calculated from unrounded assessed areas but are reported to two significant figures.

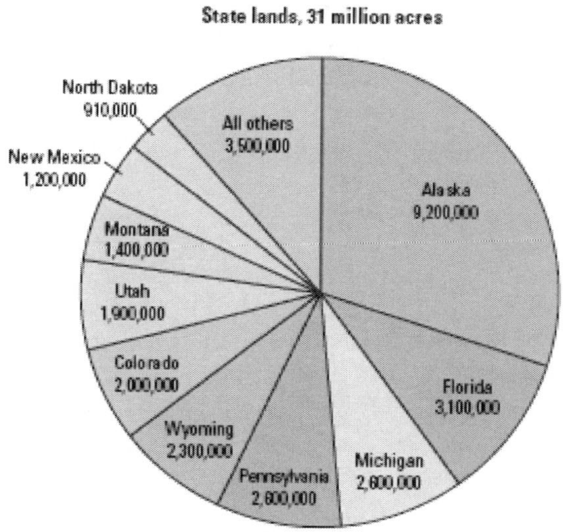

Figure 4. Pie chart showing top 10 States (and other remaining States) with the most likely estimates of State-lands allocations for land overlying areas assessed in 2012 for geologic carbon dioxide storage (U.S. Geological Survey Geologic Carbon Dioxide Storage Resources Assessment Team, 2013a, b). Resources underlying federally managed offshore areas were not assessed. Allocation areas, in acres, were calculated from unrounded assessed areas but are reported to two significant figures.

Figure 4 shows the top 10 States (and other remaining States) that have State lands overlying assessed areas; Alaska, Florida, Michigan, and Pennsylvania have the highest acreages for this particular allocation.

Whereas assessed areas are more valid for considering the geologic storage resource allocations, the TA_{SR} was tabulated for SAUs in the assessment, and understanding how lands overlie these units is also valuable for planning storage projects. Table 2*A* lists the top 10 SAUs ranked by area of overlying Federal lands. Table 2*B* lists the top 10 SAUs ranked by area of overlying BLM lands, which make up the largest Federal lands allocation and most commonly overlie assessed areas. Table 2*C* lists the top 10 SAUs ranked by area of overlying FS lands, and table 2*D* lists the top 10 SAUs ranked by area of overlying Tribal lands, the second and third largest land allocations in the assessment, respectively. Table 2*E* lists the top 10 SAUs by area of overlying State lands.

Table 2. Lists of storage assessment units (SAUs) having the largest land allocations by area and percentage for five land-management categories for U.S. areas assessed in 2012 for geologic carbon dioxide storage

[The 2012 assessment data are from the U.S. Geological Survey Geologic Carbon Dioxide Storage Resources Assessment Team (2013a). Resources underlying federally managed offshore areas were not assessed. Allocations were calculated from unrounded assessed areas but are reported to two significant figures. The top 10 SAUs having the largest total land allocations are listed for Federal lands, Bureau of Land Management (BLM) lands, U.S. Department of Agriculture Forest Service (FS) lands, Tribal lands, and State lands]

Rank	Assessed area (assessment basin code)	SAU code	SAU name	Land allocation (most likely)	
				Acres	Percent
			A. Federal lands		
1	Alaska North Slope (C5001)	C50010104	Lower Ellesmerian Deep	19,000,000	65
2	Alaska North Slope (C5001)	C50010107	Lower Torok Formation	18,000,000	73
3	Alaska North Slope (C5001)	C50010106	Beaufortian and Upper Ellesmerian	10,000,000	59
4	Williston Basin (C5031)	C50310107	Lower Swift Formation	6,600,000	14
5	Greater Green River Basin (C5037)	C50370102	Paleozoic Composite Deep	6,500,000	68
6	Williston Basin (C5031)	C50310108	Inyan Kara Group	6,200,000	14
7	U.S. Gulf Coast (C5047 and C5049)	C50490116	Tuscaloosa and Woodbine Formations	5,900,000	7.1
8	Greater Green River Basin (C5037)	C50370105	Muddy Sandstone and Cloverly Formation	5,500,000	60
9	Greater Green River Basin (C5037)	C50370109	Hilliard, Baxter, and Mancos Shales	5,300,000	63
10	Williston Basin (C5031)	C50310109	Newcastle Formation	5,300,000	15
			B. Bureau of Land Management (BLM) lands		
1	Alaska North Slope (C5001)	C50010104	Lower Ellesmerian Deep	18,000,000	64
2	Alaska North Slope (C5001)	C50010107	Lower Torok Formation	17,000,000	72
3	Alaska North Slope (C5001)	C50010106	Beaufortian and Upper Ellesmerian	10,000,000	59
4	Greater Green River Basin (C5037)	C50370102	Paleozoic Composite Deep	5,700,000	61
5	Greater Green River Basin (C5037)	C50370105	Muddy Sandstone and Cloverly Formation	4,800,000	52
6	Greater Green River Basin (C5037)	C50370109	Hilliard, Baxter, and Mancos Shales	4,800,000	56
7	Alaska North Slope (C5001)	C50010103	Lower Ellesmerian	4,700,000	52
8	Greater Green River Basin (C5037)	C50370108	Frontier Sandstone Deep	4,400,000	62
9	Greater Green River Basin (C5037)	C50370104	Nugget Sandstone Deep	3,400,000	57
10	Greater Green River Basin (C5037)	C50370107	Frontier Sandstone	3,100,000	52
			C. U.S. Department of Agriculture Forest Service (FS) lands		
1	U.S. Gulf Coast (C5047 and C5049)	C50490116	Tuscaloosa and Woodbine Formations	4,300,000	5.2
2	Williston Basin (C5031)	C50310107	Lower Swift Formation	3,700,000	7.9
3	U.S. Gulf Coast (C5047 and C5049)	C50490117	Navarro, Taylor, and Austin Groups	3,500,000	7.7
4	U.S. Gulf Coast (C5047 and C5049)	C50490104	Smackover Formation Deep	3,400,000	5.5
5	Williston Basin (C5031)	C50310108	Inyan Kara Group	3,100,000	7.1
6	Williston Basin (C5031)	C50310109	Newcastle Formation	2,700,000	8.0
7	Appalachian Basin (C5067)	C50670102	Clinton, Medina, and Tuscarora Formations	2,500,000	5.2
8	U.S. Gulf Coast (C5047 and C5049)	C50490102	Norphlet Formation Deep	2,400,000	5.5
9	U.S. Gulf Coast (C5047 and C5049)	C50490107	Sligo and Hosston Formations and Cotton Valley Group	2,200,000	4.2
10	Williston Basin (C5031)	C50310105	Kibbey Formation and Madison Group	2,200,000	6.6

Rank	Assessed area (assessment basin code)	SAU code	SAU name	Land allocation (most likely)	
				Acres	Percent
			D. Tribal lands		
1	Alaska North Slope (C5001)	C50010104	Lower Ellesmerian Deep	3,400,000	65
2	Williston Basin (C5031)	C50310107	Lower Swift Formation	3,400,000	14
3	Alaska North Slope (C5001)	C50010107	Lower Torok Formation	3,200,000	73
4	Williston Basin (C5031)	C50310101	Deadwood and Black Island Formations	3,000,000	9.1
5	San Juan Basin (C5022)	C50220102	Dakota Sandstone	2,800,000	33
6	Williston Basin (C5031)	C50310108	Inyan Kara Group	2,700,000	14
7	Williston Basin (C5031)	C50310109	Newcastle Formation	2,000,000	15
8	San Juan Basin (C5022)	C50220103	Gallup Sandstone	1,900,000	38
9	Williston Basin (C5031)	C50310105	Kibbey Formation and Madison Group	1,600,000	13
10	San Juan Basin (C5022)	C50220101	Entrada Sandstone	1,500,000	42
			E. State lands		
1	Alaska North Slope (C5001)	C50010104	Lower Ellesmerian Deep	5,800,000	20
2	Alaska North Slope (C5001)	C50010106	Beaufortian and Upper Ellesmerian	4,900,000	28
3	Alaska North Slope (C5001)	C50010108	Upper Torok Formation	3,300,000	89
4	Alaska North Slope (C5001)	C50010103	Lower Ellesmerian	2,900,000	32
5	Appalachian Basin (C5067)	C50670102	Clinton, Medina, and Tuscarora Formations	2,900,000	5.9
6	Alaska North Slope (C5001)	C50010107	Lower Torok Formation	2,800,000	12
7	Appalachian Basin (C5067)	C50670104	Oriskany Sandstone	2,700,000	7.1
8	Michigan Basin (C5063)	C50630101	Ordovician and Cambrian Composite	2,600,000	8.6
9	South Florida Basin (C5050)	C50500105	Cedar Keys and Lawson Formations	2,300,000	10
10	Michigan Basin (C5063)	C50630102	Salina Group and Middle Silurian Composite	2,300,000	10

There is commonality when comparing the top 10 SAUs in table 2*A* by area and the 10 SAUs having the largest TA_{SR} listed by the U.S. Geological Survey Geologic Carbon Dioxide Storage Resources Assessment Team (2013b, p. 15 and table 3). This commonality includes the Lower Torok Formation SAU (C50010107 with a TA_{SR} of 140 Gt), in the Alaska North Slope study area, ranked fifth largest in the 2012 assessment and ranked second in this table of SAUs with the most Federal lands (about 18 million acres or about 27,000 square miles). Additional overlap between the lists includes the Tuscaloosa and Woodbine Formations SAU (C50490116 with a TA_{SR} of 85 Gt), in the U.S. Gulf Coast, ranked eighth in storage capacity and seventh on the list of SAUs with the most Federal lands (about 5.9 million acres or about 9,100 square miles).

Conclusion

This report has estimates of acreages in five landownership categories (Federal lands, State lands, Tribal lands, private and other lands, and offshore areas) for lands that overlie areas assessed by the USGS in 2012 for geologic carbon dioxide (CO_2) storage (U.S. Geological Survey Geologic Carbon

Dioxide Storage Resources Assessment Team, 2013b). The Bureau of Land Management (BLM) is the Federal agency with the most land (about 80 million acres, or about 126,000 square miles) overlying assessed areas and storage assessment units (SAUs), followed by the U.S. Department of Agriculture Forest Service (FS), which has about 27 million acres (or about 42,000 square miles). Also significantly, Tribal lands cover about 17 million acres (about 26,000 square miles). Alaska and Nevada are the States with the most Federal lands overlying assessed areas, which are mostly managed by the BLM. The Lower Torok Formation SAU in the Alaska North Slope and the Tuscaloosa and Woodbine Formations SAU in the U.S. Gulf Coast are examples of units having large amounts of technically accessible storage resources (TA_{SR}) underlying Federal lands. The Alaska Region and the Rocky Mountains and Northern Great Plains Region contain the most storage resources area overlain by Federal lands, whereas the Coastal Plains Region contains the most TA_{SR} regardless of land ownership.

Although officials at individual Federal sites have estimated the mineral rights, the USGS did not distinguish between mineral rights and pore-space allocation for the geologic CO_2 storage resources assessment. This report should not be used to define title to any specific area because of the complicated nature of surface and subsurface mineral-estate ownership. The scattered geographic distribution of the SAUs and assessed areas, and the existence of legal and surface restrictions on the accessibility of the TA_{SR} could eliminate much of the estimated resources from consideration for development. Nevertheless, information on Federal management of storage resources in this report and other results included in the 2013 assessment report will assist policymakers and planners in making decisions concerning the multiple uses of Federal lands and the scope of future geologic CO_2 storage development.

Additional Assessment Information

Assessment results, supporting geologic studies of the storage formations and regional seals composing the storage assessment units, and the methodology used in assessing the national geologic carbon dioxide storage resource are available online at the USGS Energy Resources Program Web site http://energy.usgs.gov/EnvironmentalAspects/EnvironmentalAspectsofEnergyProductionandUse/GeologicCO2Sequestration.aspx or more simply at http://go.usa.gov/8X8.

GLOSSARY

Federal lands One of five land-ownership categories used in this assessment for allocation of resources. Federal lands are lands within the United States owned or administered by the Federal Government. These include parks, national wildlife refuges, military reservations, Federal prisons, and public-domain land. Spatial data for this category came from State-specific spatial resources from the Bureau of Land Management (BLM) available at http://www.blm.gov and from the National Atlas of the United States (U.S. Geological Survey, 2005a).

federally owned offshore areas Federal jurisdiction begins at 3 geographic (nautical) miles from the established baseline for the coast and extends to an outer limit of 200 nautical miles (Outer Continental Shelf). However, there are special cases. Because of claims existing at the dates of statehood, Texas and the Gulf Coast of Florida have proprietary interest in a submerged belt of land, 9 geographic miles wide, extending seaward along the coast (Thormahlen, 1999). Resource assessments in federally owned offshore areas are typically done by the Bureau of Ocean Energy Management (BOEM).

offshore areas One of five land-ownership categories used in this assessment for allocation of resources. In this assessment, offshore areas include State waters (see definition), and some are managed by Federal agencies, such as the BLM. Both State and Federal offshore area boundaries came from the National Atlas of the United States (Minerals Management Service, 2005).

private and other lands One of five land-ownership categories used in this assessment for allocation of resources. Lands not owned by Federal, State, or Tribal entities are placed in this category of private and other lands. These lands either belong to private owners or, less commonly, belong to additional BLM-named ownership categories (such as county, city, water, and other) and are grouped together here. Spatial data for this category came primarily from State-specific spatial resources from the BLM available at http://www.blm.gov; when data were not available from the BLM, they were obtained from individual State agencies.

State lands One of five land-ownership categories used in this assessment for allocation of resources. This ownership category includes lands owned by State entities, as categorized by the BLM in State-specific land-ownership data obtained from http://www.blm.gov; when data were not available from the BLM, they were obtained from individual State agencies.

State waters State jurisdiction begins at the established baseline for the coast and extends 3 geographic (nautical) miles. However, there are special cases. Because of claims existing at the dates of statehood, Texas and the Gulf Coast of Florida have proprietary interest in a submerged belt of land, 9 geographic miles wide, extending seaward along the coast (Thormahlen, 1999).

technically accessible storage resource (TA_{SR}) The mass of carbon dioxide (CO_2) that may be injected and stored using present-day geologic and hydrologic knowledge of the subsurface and engineering practices. This term is analogous to the term "technically recoverable resource" used in U.S. Geological Survey oil and gas assessments.

Tribal lands One of five land-ownership categories used in this assessment for allocation of resources. Indian or Tribal lands within the United States are areas with boundaries established by treaty, statute, and (or) executive or court order, recognized by the Federal Government as territory in which American Indian Tribes have primary governmental authority. Spatial data for this category were revised slightly from data in U.S. Geological Survey Geologic Carbon Dioxide Storage Resources Assessment Team (2013a) and come from State-specific spatial resources from the BLM available at http://www.blm.gov and from the National Atlas of the United States (U.S. Geological Survey, 2005b).

REFERENCES

Blondes, M.S., Brennan, S.T., Merrill, M.D., Buursink, M.L., Warwick, P.D., Cahan, S.M., Cook, T.A., Corum, M.D., Craddock, W.H., DeVera, C.A., Drake, R.M., II, Drew, L.J., Freeman, P.A., Lohr, C.D., Olea, R.A., Roberts-Ashby, T.L., Slucher, E.R., and Varela, B.A., 2013, National assessment of geologic carbon dioxide storage resources— Methodology implementation: U.S. Geological Survey Open-File Report 2013–1055, 26 p., http://pubs.usgs.gov/ of/2013/1055/.

Brennan, S.T., Burruss, R.C., Merrill, M.D., Freeman, P.A., and Ruppert, L.F., 2010, A probabilistic assessment methodology for the evaluation of geologic carbon dioxide storage: U.S. Geological Survey Open-File Report 2010–1127, 31 p., http://pubs.usgs.gov/of/2010/1127.

Burruss, R.C., Brennan, S.T., Freeman, P.A., Merrill, M.D., Ruppert, L.F., Becker, M.F., Herkelrath, W.N., Kharaka, Y.K., Neuzil, C.E., Swanson, S.M., Cook, T.A., Klett, T.R., Nelson, P.H., and Schenk, C.J., 2009,

Development of a probabilistic assessment methodology for evaluation of carbon dioxide storage: U.S. Geological Survey Open-File Report 2009–1035, 81 p., http://pubs.usgs.gov/ of/2009/1035/.

Dolence, R.C., Kirschner, Charles, Babel, Christopher, Findle, Patrick, and Harouff, Steven, 2009, Storage of captured carbon dioxide beneath Federal lands: U.S. Department of Energy, National Energy Technology Laboratory Report DOE/NETL–2009/1358, 75 p., accessed June 23, 2014, at http://www.nctl.doe.gov/ File%20Library/Research/Energy%20Analysis/Coal/ Fed-Land_403-01-02_050809.pdf.

Karlsen, A.W., SanFilipo, J.R., and Warwick, P.D., 2002, Coal under lying Federal lands in the Gulf of Mexico Coastal Plain: U.S. Geological Survey Fact Sheet 094–02, 2 p., http://pubs.usgs.gov/fs/fs094-02/.

Minerals Management Service, 2005, Official protraction diagram and leasing map boundaries: National Atlas of the United States, map layer accessed in 2005 at http://nationalatlas.gov/mld/opdlm0p.html. [After the National Atlas of the United States was retired in 2014, some of its products were made available at http://nationalmap.gov/small_scale/.]

Society of Petroleum Engineers, 2014, Carbon capture and sequestration: Society of Petroleum Engineers Carbon Capture and Sequestration Committee Web site, accessed September 11, 2014, at http://www.spe.org/industry/sites/ccs/.

Thormahlen, L.F., 1999, Boundary development on the Outer Continental Shelf: Minerals Management Service, Mapping and Boundary Branch OCS report MMS 99–0006, 12 p., accessed February 21, 2013, at http://www. boem.gov/BOEM- Newsroom/Library/Publications/1999/99-0006-pdf.aspx.

U.S. Geological Survey, 2005a, Federal lands of the United States: National Atlas of the United States, map layer accessed in 2005 at http://www. nationalatlas.gov/mld/ fedlanp.html. [After the National Atlas of the United States was retired in 2014, some of its products were made available at http://nationalmap.gov/small_scale/.]

U.S. Geological Survey, 2005b, Indian lands of the United States: National Atlas of the United States, map layer accessed in 2005 at http://nationalatlas.gov/mld/indlanp. html. [After the National Atlas of the United States was retired in 2014, some of its products were made available at http://nationalmap.gov/small_scale/.]

U.S. Geological Survey Geologic Carbon Dioxide Storage Resources Assessment Team, 2013a, National assessment of geologic carbon dioxide storage resources—Data (ver. 1.1, September 2013): U.S. Geological Survey Data

Series 774, 13 p., plus 2 appendixes and 2 large tables in separate files, http://pubs.usgs.gov/ds/774/.

U.S. Geological Survey Geologic Carbon Dioxide Storage Resources Assessment Team, 2013b, National assessment of geologic carbon dioxide storage resources—Results (ver. 1.1, September 2013): U.S. Geological Survey Circular 1386, 41 p., http://pubs.usgs.gov/circ/1386/.

U.S. Geological Survey Geologic Carbon Dioxide Storage Resources Assessment Team, 2013c, National assessment of geologic carbon dioxide storage resources—Summary (ver. 1.1, September 2013): U.S. Geological Survey Fact Sheet 2013–3020, 6 p., http://pubs.usgs.gov/fs/2013/3020/.

Warwick, P.D., and Corum, M.D., eds., 2012, Geologic framework for the national assessment of carbon dioxide storage resources: U.S. Geological Survey Open-File Report 2012–1024, http://dx.doi.org/10.3133/ofr20121024. [Chapters A–J were posted by November 1, 2014.]

INDEX

A

B

C

D

E

F

G

H

I

J

L

M

N

O

P